20

TESI

THESES

tesi di perfezionamento in Matematica sostenuta il 15 marzo 2011

COMMISSIONE GIUDICATRICE
Luigi Ambrosio, Presidente
Marco Abate
Filippo Bracci
Charles Favre
Pietro Majer
Giorgio Patrizio
Giuseppe Tomassini

Matteo Ruggiero
IMJ - Université Paris Diderot 75205 Paris Cedex 13, France

Rigid Germs, the Valutative Tree, and Applications to Kato Varieties

Matteo Ruggiero

Rigid Germs, the Valuative Tree, and Applications to Kato Varieties

EDIZIONI
DELLA
NORMALE

ISBN 978-88-7642-558-5
e-ISBN 978-88-7642-559-2

a Laura e zia Anna

Contents

Introduction

Holomorphic dynamics has several points of view: it can be discrete or continuous, and be studied locally or globally, but all these aspects are, sometimes surprisingly and in a very fascinating way, linked to one another.

The setting of global discrete holomorphic dynamics is the following: one has a complex space X of dimension d, and a holomorphic map $f : X \to X$, and wants to understand the behavior of the iterates $f^{\circ n}$ of f. For example one can check if the orbit of a point $x \in X$ (*i.e.*, the set $\{f^{\circ n}(x) \mid n \in \mathbb{N}\}$) changes regularly by moving the starting point x.

On the other hand, local discrete holomorphic dynamics still studies the behavior of a map f, but near a given fixed point p, and hence in coordinates one is interested into the behavior of a holomorphic germ $f : (\mathbb{C}^d, 0) \to (\mathbb{C}^d, 0)$ and its iterates, existence of basins of attractions, or the structure of the stable set (where all the iterates of f are defined in a neighborhood of 0).

One of the main techniques to study the dynamics of a family $\mathcal{F}$ of holomorphic germs is looking for *normal forms*. Roughly speaking, one looks for a (possibly small) family $\mathcal{G}$ of germs, whose dynamics is easier to study, and such that every $f \in \mathcal{F}$ can be reduced to a germ $g \in \mathcal{G}$ by changing coordinates.

Definition. Let $f, g : (\mathbb{C}^d, 0) \to (\mathbb{C}^d, 0)$ be two holomorphic germs. We shall say that f and g are (*holomorphically*, *topologically*, *formally*) *conjugated* if there exists a (biholomorphism, homeomorphism, formal invertible map) $\phi : (\mathbb{C}^d, 0) \to (\mathbb{C}^d, 0)$ such that

$$\phi \circ f = g \circ \phi.$$

Depending on the regularity of the change of coordinates: holomorphic, homeomorphic, formal, we talk about holomorphic, topological or formal classification.

We can say that holomorphic dynamics was born in 1884, when Kœnigs (in [44]) proved a conjugacy result in local discrete dynamics in dimension $d = 1$.

Theorem (Kœnigs). *Let $f : (\mathbb{C}, 0) \to (\mathbb{C}, 0)$ be a holomorphic germ such that the multiplier $\lambda := f'(0)$ is such that $|\lambda| \neq 0, 1$. Then f is holomorphically conjugated to the linear part $z \mapsto \lambda z$.*

Twenty years later, Böttcher proved a result on the same lines for non-invertible germs (see [10]).

Theorem (Böttcher). *Let $f : (\mathbb{C}, 0) \to (\mathbb{C}, 0)$ be a holomorphic germ of the form*

$$f(z) = a_p z^p + a_{p+1} z^{p+1} + \dots,$$

with $p \geq 2$ and $a_p \neq 0$. Then f is holomorphically conjugated to the map $z \mapsto z^p$.

Still in the beginning of the 20th century, Leau (see [46] and [47]) and Fatou (see [24]) proved a local conjugacy result for the parabolic case, *i.e.*, when $f : (\mathbb{C}, 0) \to (\mathbb{C}, 0)$ is such that its multiplier $\lambda = f'(0)$ is a root of 1. Up to taking a suitable iterate of f, we can suppose that $\lambda = 1$.

Definition. Let $f : (\mathbb{C}^d, 0) \to (\mathbb{C}^d, 0)$ be a holomorphic germ. It is called *tangent to the identity* if $df_0 = \mathrm{Id}$.

In dimension $d = 1$, write $f : (\mathbb{C}, 0) \to (\mathbb{C}, 0)$ in the form

$$f(z) = z(1 + a_k z^k + a_{k+1} z^{k+1} + \dots),$$

with $k \geq 1$ and $a_k \neq 0$. Then $k + 1$ is called the *parabolic multiplicity* of f.

Definition. A *parabolic domain* in $\mathbb{C}$ is a simply connected open domain Δ such that $0 \in \partial\Delta$.

A parabolic domain Δ is said to be an *attracting petal* (resp., *repelling petal*) for a map f tangent to the identity if $f(\Delta) \subset \Delta$ and $f^{\circ n}(x) \to 0$ for every $x \in \Delta$ (resp., the same for f^{-1}).

Theorem (Leau, Fatou). *Let $f : (\mathbb{C}, 0) \to (\mathbb{C}, 0)$ be a tangent to the identity germ with parabolic multiplicity $k+1$. Then there exist k attracting petals and k repelling petals such that in every petal f is holomorphically conjugated to $z \mapsto z+1$. Distinct attracting petals are disjoint, and the same holds for repelling petals. The union of attracting and repelling petals form a punctured neighborhood of 0.*

The formal and topological classifications for this kind of germs are not so difficult at least to state, but the holomorphic classification is surprisingly complicated: the moduli space is infinite-dimensional, and the final answer of this question was given almost 70 years later, independently by Écalle using resurgence theory (see [21, 22, 23]) and Voronin (see [61]).

Fatou, Julia, Cremer, Siegel, Brjuno, Sullivan, Douady, Hubbard, Yoccoz and many others gave their contribution to the study of holomorphic dynamics in dimension 1, and right now most of the main issues for both local and global holomorphic dynamics in dimension 1 are solved.

In higher dimensions, a very fruitful theory has been developed for the global setting, by Bedford, Sibony, Fornaess, Smillie and many others, whereas in the local setting only a few simpler cases are well understood, such as the invertible attracting case, while the more complicated ones are still subject of study, even in dimension $d = 2$.

Definition. Let $f : (\mathbb{C}^2, 0) \to (\mathbb{C}^2, 0)$ be a holomorphic germ, and let us denote by $\mathrm{Spec}(df_0) = \{\lambda_1, \lambda_2\}$ the set of eigenvalues of df_0. Then f is said:

- *attracting* if $|\lambda_i| < 1$ for $i = 1, 2$;
- *superattracting* if $df_0 = 0$;
- *nilpotent* if df_0 is nilpotent (*i.e.*, $df_0^2 = 0$; in particular, superattracting germs are nilpotent germs);
- *semi-superattracting* if $\mathrm{Spec}(df_0) = \{0, \lambda\}$, with $\lambda \neq 0$;
- *of type* $(0, D)$ if $\mathrm{Spec}(df_0) = \{0, \lambda\}$ and $\lambda \in D$, with $D \subset \mathbb{C}$ a subset of the complex plane.

In particular the semi-superattracting germs are the ones of type $(0, \mathbb{C}^*)$.

We shall always consider only *dominant* holomorphic germs, *i.e.*, holomorphic germs f such that $\det df_z \not\equiv 0$. For non-dominant holomorphic germs, the dynamics is essentially 1-dimensional.

Favre in 2000 gave the holomorphic classification of a special type of germs, namely the (attracting) *rigid germs* (see [25]).

Definition. Let $f : (\mathbb{C}^d, 0) \to (\mathbb{C}^d, 0)$ be a holomorphic germ. We denote by $\mathcal{C}(f) = \{z \mid \det(df_z) = 0\}$ the *critical set* of f, and by $\mathcal{C}(f^\infty) = \bigcup_{n\in\mathbb{N}} f^{-n}\mathcal{C}(f)$ the *generalized critical set* of f. Then a (dominant) holomorphic germ f is *rigid* if:

(i) $\mathcal{C}(f^\infty)$ (is empty or) has simple normal crossings (SNC) at the origin; and
(ii) $\mathcal{C}(f^\infty)$ is forward f-invariant.

Another very interesting class of holomorphic germs is given by strict germs, that (in dimension 2, but not in higher dimensions) are a subset of rigid germs.

Definition. Let $f : (\mathbb{C}^d, 0) \to (\mathbb{C}^d, 0)$ be a (dominant) holomorphic germ. Then f is a **strict germ** if there exist a SNC divisor with support C and a neighborhood U of 0 such that $f|_{U \setminus C}$ is a biholomorphism with its image.

Besides giving interesting classes of examples of local dynamics in higher dimensions, in the 2-dimensional case rigid and strict germs are very important for at least two reasons: first, every (dominant) holomorphic germ is birationally conjugated to a rigid germ; second, every strict germ gives rise to a compact complex non-Kähler surface (Kato surface).

Valuative tree

A very useful tool for the study of holomorphic dynamics (in dimension 2 or higher) has been borrowed from algebraic geometry: *blow-ups*. Roughly speaking, a blow-up of a point p in $\mathbb{C}^d$ consists in replacing p by the set $\mathbb{P}(T_p X)$, *i.e.*, the set of "directions" through p.

If then one wants to study holomorphic dynamics locally at $0 \in \mathbb{C}^2$, one can look for a suitable *modification* over 0, (*i.e.*, a sequence of blow-ups, the first one over 0), to get a simpler dynamical situation on the blown-up space.

These techniques were first used for studying foliations by Seidenberg, Camacho and Sad, and many others (see, *e.g.*, [55] and [13]), and then transferred, by Hakim, Abate and others, to the tangent to the identity case in local dynamics in $\mathbb{C}^2$ (see [34] and [2]).

To study local (and global) holomorphic dynamics, Favre and Jonsson in [26] developed a tool, the *valuative tree*, that roughly speaking is a way to look at all possible modifications over the origin.

Using the valuative tree, and the action induced on it by a germ $f : (\mathbb{C}^2, 0) \to (\mathbb{C}^2, 0)$, they proved that up to modifications you can suppose that a *super-attracting* germ is actually rigid. Let us be more precise.

Definition. Let $f : (\mathbb{C}^2, 0) \to (\mathbb{C}^2, 0)$ be a (dominant) holomorphic germ. Let $\pi : X \to (\mathbb{C}^2, 0)$ be a modification and $p \in \pi^{-1}(0)$ a point in the exceptional divisor of π. Then we shall call the triple $(\pi, p, \hat{f})$ a *rigidification* for f if the lift $\hat{f} = \pi^{-1} \circ f \circ \pi$ is a holomorphic rigid germ with fixed point $p = \hat{f}(p)$.

Finding a rigidification is, a priori, extremely hard, since if we have a germ $f : (\mathbb{C}^2, 0) \to (\mathbb{C}^2, 0)$, a modification $\pi : X \to (\mathbb{C}^2, 0)$ and a point $p \in \pi^{-1}(0)$, the lift $\hat{f} = \pi^{-1} \circ f \circ \pi$ in general is just a rational map, and it is already not easy to have $\hat{f}$ to be a holomorphic germ in a fixed point p.

In this thesis we extend Favre's and Jonsson's result (see [27, Theorem 5.1]) to germs with non-invertible differential in 0 (for invertible germs, the result is trivial, being the map already rigid), getting

Theorem. *Every (dominant) holomorphic germ $f : (\mathbb{C}^2, 0) \to (\mathbb{C}^2, 0)$ admits a rigidification.*

Then we shall study more carefully the case of *semi-superattracting* germs, getting a sort of uniqueness of the rigidification process, and a result on the existence (and uniqueness) of invariant curves.

Theorem. *Let f be a (dominant) semi-superattracting holomorphic germ of type $(0, \lambda)$. Then there exist two curves C and D such that the following holds:*

- *C is a (possibly formal) curve through 0, with multiplicity equal to 1 and tangent to the λ-eigenspace of df_0, such that $f(C) = C$.*
- *D is a (holomorphic) curve through 0, with multiplicity equal to 1 and tangent to the 0-eigenspace of df_0, such that either $f(D) = D$ or $f(D) = 0$.*
- *There are no other invariant or contracted (not even formal) curves for f besides C and D.*

Thanks to this result, the formal classification of semi-superattracting rigid germs can be found. We shall quote here only a consequence of this classification (see Section 2.5 for the precise statement):

Corollary. *The moduli space of semi-superattracting germs in $\mathbb{C}^2$ up to formal conjugacy is infinite-dimensional.*

This result shows how difficult (if not impossible) is to give an explicit classification of semi-superattracting germs up to holomorphic conjugacy. As a matter of fact, one has always to consider the complexity of the generalized critical set, that generally has an infinite number of irreducible components.

With the rigidification result and the last remark in mind, we can then focus on better understanding the dynamics of rigid germs. For semi-superattracting rigid germs of type $(0, \mathbb{D})$ Favre's result gives the holomorphic classification; in this thesis we focus our attention on a sort of limit case, germs of type $(0, 1)$. Hakim proved (see [33]) the following result on the existence of basins of attraction.

Definition. Let $f : (\mathbb{C}^2, 0) \to (\mathbb{C}^2, 0)$ be a holomorphic germ of type $(0, 1)$. Let C be the f-invariant (formal) curve associated to the

1-eigenspace of the differential df_0 at 0, parametrized by a suitable (formal) map $\gamma : \mathbb{C}[[t]] \to \mathbb{C}[[z, w]]$. Then we shall call *parabolic multiplicity* of f the parabolic multiplicity of $\gamma^{-1} \circ f|_C \circ \gamma$.

Theorem (Hakim). *Let $f : (\mathbb{C}^2, 0) \to (\mathbb{C}^2, 0)$ be a holomorphic germ of type $(0, 1)$, with parabolic multiplicity $k + 1$, and let us denote by $\mathbb{D}_\rho$ the open disc in $\mathbb{C}$ centered at 0 and with radius $\rho > 0$. Then there exist k (disjoint) parabolic domains $\Delta_0, \ldots, \Delta_{k-1} \subset \mathbb{C}$, such that, for ρ small enough,*

$$W_j := \Delta_j \times \mathbb{D}_\rho$$

are basins of attraction for f and there exist holomorphic submersions

$$\phi_j : W_j \to \mathbb{C}$$

that satisfy the following functional equation:

$$\phi_j\big(f(p)\big) = \phi_j(p) + 1.$$

Notice that, even if the basins of attraction are a product of a parabolic domain Δ_j and a disc $\mathbb{D}_\rho$, the parabolic domain $\Delta_j \times \{0\}$ is not necessarily f-invariant (as happens for germs tangent to the identity, see [34]), and f might not admit parabolic curves.

Definition. A *parabolic curve* for a germ $f : (\mathbb{C}^2, 0) \to (\mathbb{C}^2, 0)$ at the origin is a injective holomorphic map $\varphi : \Delta \to \mathbb{C}^d$ satisfying the following properties:

(i) Δ is a parabolic domain in $\mathbb{C}$;
(ii) φ is continuous at the origin, and $\varphi(0) = 0$;
(iii) $\varphi(\Delta)$ is invariant under f, and $f^{\circ n}(z) \to 0$ for every $z \in \varphi(\Delta)$.

Roughly speaking, Hakim's result tells us the behavior of "one coordinate" of f in a basin of attraction. We can focus on understanding the behavior of f also with respect to the "other coordinate", to get a complete description of the dynamics f in the basins of attraction. For the reasons we anticipated above, we shall consider rigid germs and prove the following result.

Theorem. *Let $f : (\mathbb{C}^2, 0) \to (\mathbb{C}^2, 0)$ be a holomorphic rigid germ of type $(0, 1)$ of parabolic multiplicity $k + 1$. Let*

$$W_j := \Delta_j \times \mathbb{D}_\rho$$

for $j = 0, \dots, k-1$ *be basins of attraction for* f *as above. If there is a parabolic curve in* W_j*, then there exists a holomorphic conjugation* $\Phi_j : W_j \to \widetilde{W}_j$ *between* $f|_{W_j}$ *and the map*

$$\widetilde{f}(z, w) = \left(\frac{z}{\sqrt[k]{1+z^k}}, z^c w^d \big(1 + \widetilde{h}(z)\big) \right),$$

where $\widetilde{W}_j$ *is a suitable parabolic domain. Moreover, if* $d \geq 2$ *then we can get* $\widetilde{h} \equiv 0$.

In particular, the action of such germ in the second coordinate is either monomial or linear with respect to w.

The assumption of the existence of parabolic curves is quite technical; the feeling on the matter is that, if parabolic curves do not exist, then f should be conjugated (in each basin of attraction) to a map that in the second coordinate is affine with respect to w.

Kato Varieties

Coming back to local dynamics in the attracting case, given an attracting germ $f : (\mathbb{C}^d, 0) \to (\mathbb{C}^d, 0)$, it is natural to look at its basin of attraction to 0, and its fundamental domain, *i.e.*, (any dense open subset of) the basin of attraction modulo the action of f itself.

Starting from the 70's, Kato, Inoue, Dloussky, Oeljeklaus, Toma and others proved that one can find some compactifications of these fundamental domains when $f : (\mathbb{C}^2, 0) \to (\mathbb{C}^2, 0)$ is also strict, getting interesting examples of compact complex non-Kähler surfaces, called *Kato surfaces*.

Kato surfaces are of great interest also for the Kodaira-Enriques classification of compact complex surfaces. Indeed, they all belong to the so called *Class* VII.

Definition. A surface X is called of *class* VII if it has Kodaira dimension $\operatorname{kod}(X) = -\infty$ and first Betti number $b_1 = 1$. If moreover X is a minimal model, it is called of *class* VII_0.

We have to explain now what is the Kodaira dimension of a compact complex manifold, and what is a minimal model. We shall start from the second.

Definition. A compact complex surface X is called *minimal model* if it does not exist a compact complex surface Y and a modification $\pi : X \to Y$.

This definition can seem difficult to check directly, but thanks to the Castelnuovo-Enriques criterion (see [31, p. 476]), it is equivalent to asking that the surface X has no *exceptional curves*, *i.e.*, rational curves with self-intersection -1.

Theorem (Castelnuovo-Enriques Criterion). *Let X be a 2-manifold, and $D \subset X$ a curve in X. Then there exists a 2-manifold Y so that $\pi : X \to Y$ is the blow-up of a point $p \in Y$ with $D = \pi^{-1}(p)$ if and only if D is a rational curve of self-intersection -1.*

So, up to modifications (and hence up to birationally equivalent models), a compact complex surface can be supposed to be a minimal model.

If reducing to minimal models can be considered as the first step for the classification of compact complex surfaces, the second step would be sorting surfaces with respect to the *Kodaira dimension*.

Definition. Let X be a compact complex n-manifold. For every $m \in \mathbb{N}^*$, we shall call the m-th *plurigenera* the dimension

$$P_m := h^0\big(X, \mathcal{O}(m\mathcal{K}_X)\big)$$

of the space of holomorphic sections of the line bundle $m\mathcal{K}_X$, where $\mathcal{K}_X = \bigwedge^n T^*X$ is the *canonical bundle* of X (here T^*X denotes the holomorphic cotangent bundle of X).

The *Kodaira dimension* of X is

$$\operatorname{kod}(X) = \min\{k \mid P_m = O(m^k) \text{ for } m \to +\infty\}.$$

The Kodaira dimension somehow tells us how positive is the canonical bundle.

When $P_m = 0$ for every $m \geq 1$, we shall say that the Kodaira dimension is $-\infty$.

For a (compact complex) n-manifold, the Kodaira dimension can take values in $\{-\infty, 0, \dots, n\}$. The case $\operatorname{kod}(X) = n$ is said to be of *general type*.

In dimension 2, surfaces of general type are not completely understood. There are results on the structure of the moduli spaces, but it seems not easy to compute them for all cases.

However for Kodaira dimension 1 and 0 the classification is done and classical, while for Kodaira dimension $-\infty$, only one case is still not completely understood: class VII surfaces.

When the second Betti number $b_2(X) = 0$, these surfaces have been completely classified, thanks to the work of Kodaira ([42, 43]), Inoue ([38]), Bogomolov ([8]), Li, Yau and Zheng ([48]), Teleman ([60]).

For $b_2 > 0$, the classification is not completed yet. Before describing the known results, we need a definition.

Definition. Let X be a compact complex n-manifold. A *spherical shell* is a holomorphic embedding $i : V \hookrightarrow X$, where V is a neighborhood of $\mathbb{S}^{2n-1} = \partial\mathbb{B}^{2n} \subset \mathbb{C}^n$. A spherical shell is said *global* (or *GSS*) if $X \setminus i(V)$ is connected.

Kato introduced a construction method for surfaces of class VII_0 with $b_2 > 0$, called *Kato surfaces* (see [40]), that starts from a Kato data.

Definition. Let $B = \overline{\mathcal{B}_\varepsilon}$ be a closed ball in $\mathbb{C}^n$ of center 0 and radius $\varepsilon > 0$, and $\pi : \widetilde{B} \to B$ a modification over 0. Let $\sigma : B \to \widetilde{B}$ be a biholomorphism with its image such that $\sigma(0)$ is a point of the exceptional divisor of π. The couple (π, σ) is called a *Kato data.*

Kato datas and (rigid, strict) germs are strictly related, as the following definition shows.

Definition. Let (π, σ) be a Kato data. Then we can consider $f_0 = \pi \circ \sigma : B \to B$, that turns out to be a holomorphic rigid and strict germ, with a fixed point in 0 the center of B. We shall call this germ the *base germ* associated to the given Kato data.

On the other hand, given a (rigid and strict) holomorphic germ $f_0 : (\mathbb{C}^n, 0) \to (\mathbb{C}^n, 0)$, we shall call *resolution* for f_0 a decomposition $f_0 = \pi \circ \sigma$, with π a modification over 0 and σ a (germ) biholomorphism that sends 0 into a point of the exceptional divisor of π.

Then, roughly speaking, a Kato variety is constructed as follows.

Definition. Let (π, σ) be a Kato data. Let $B = \mathcal{B}_\varepsilon$ be a ball in $\mathbb{C}^n$ of center 0 and radius $\varepsilon > 0$. The *Kato variety* associated to the given Kato data is the quotient of $\widetilde{X} = \pi^{-1}(\overline{B}) \backslash \sigma(B)$ by the action of $\sigma \circ \pi : \pi^{-1}(\partial B) \to \sigma(\partial B)$, that is a biholomorphism on a suitable neighborhood of $\pi^{-1}(\partial B)$ for ε small enough.

Dloussky in his PhD thesis [15] studied deeply this construction and properties of Kato surfaces. Among these properties, we shall underline the following (see [40] and [20]).

Theorem (Kato, Dloussky-Oeljeklaus-Toma). *Let X be a surface of class VII_0 with $b_2 = b_2(X) > 0$. Then X admits at most b_2 rational curves. Moreover, X admits a GSS if and only if X has exactly b_2 rational curves.*

There are no known examples of surfaces X of class VII_0 that do not admit global spherical shells, and a big conjecture (called the GSS Conjecture) claims that there are none.

Dloussky and Oeljeklaus (see [17]) studied the case when the germ f_0 arises from the action at infinity of an automorphism of $\mathbb{C}^2$.

Definition. An automorphism $f : \mathbb{C}^2 \to \mathbb{C}^2$ is said to be a *Hénon map* if it is of the form

$$f(x, y) = \big(p(x) - ay, x\big),$$

with p a polynomial of degree $d = \deg p \geq 2$.

Polynomial automorphisms of $\mathbb{C}^2$ can be subdivided into two classes, *elementary automorphisms*, whose dynamics is easier to study, and compositions of Hénon maps; see [29].

The idea is then to consider the extension $F : \mathbb{P}^2 \to \mathbb{P}^2$, that has an indeterminacy point at $[0 : 1 : 0]$, and an indeterminacy point for the inverse at $[1 : 0 : 0]$.

Looking at the action of F on the line at infinity, one finds that there is a fixed point $p = [1 : 0 : 0]$, and the germ $F_p := f_0$ is strict and admits a resolution, and hence an associated Kato surface.

This new approach gives a connection between the dynamics of f and the Kato surface X associated to f_0: in particular, if we denote by U the basin of attraction of f to p, then X turns out to be a compactification of the fundamental domain V of U.

From this dynamical interpretation of Kato surfaces, some questions arise:

- Can we add a point "at infinity" to V and get a (possibly singular) compact complex manifold? Or equivalently, is the Alexandroff one-point compactification of V a (possibly singular) complex manifold?
- Can we lift objects that are invariant for f_0 to X, obtaining some additional structure on X (such as the existence of subvarieties, foliations, vector fields)?

The first property is actually equivalent to contracting all the rational curves to a point, that was a result already known for Kato surfaces (see [15]).

The second phenomenon has been studied for example by Dloussky, Oeljeklaus and Toma, starting from [15]; see, *e.g.*, [16, 18] and [19].

Favre, in his PhD thesis (see [25]), used his classification of attracting rigid germs in $\mathbb{C}^2$ and studied the construction of a holomorphic foliation on X, and the computation of the first fundamental group of the basin U, already obtained using other techniques by Hubbard and Oberste-Vorth (see [37]), that turns out to be very complicated.

In this work, we shall study an example of these phenomena in dimension 3. Several aspects become more complicated. First of all, while the structure of polynomial automorphisms in $\mathbb{C}^2$ is pretty clear, in higher dimensions it is still a subject of research, and we have no "Hénon maps" that we can use. Indeed only a few cases have been classified: for instance, the automorphisms of degree 2 in $\mathbb{C}^3$ (see [30] and [49]).

However, Sibony and others identified a special property of Hénon maps, *regularity*, and studied maps with this property in higher dimensions, getting a very fruitful theory on global holomorphic dynamics, using currents and pluripotential theory.

Definition. Let $f : \mathbb{C}^n \to \mathbb{C}^n$ be a polynomial automorphism, $F : \mathbb{P}^n \to \mathbb{P}^n$ be its extension to $\mathbb{P}^n$, and denote by I^+ and I^- the indeterminacy sets for F and F^{-1} respectively. Then f is said to be *regular* (in the sense of Sibony) if $I^+ \cap I^- = \emptyset$.

Furthermore, the structure of birational maps is more complicated in dimension ≥ 3: one can blow-up not only points, but also curves and varieties of higher dimension, and not every birational map is obtained as a composition of blow-ups followed by a composition of blow-downs (the inverse of blow-ups, see [9]). Moreover, the problem of finding an equivalent of minimal models in dimensions higher than 2 is still open (the project to solve this problem is called "Minimal Model Program", based on Mori theory).

So not so much is known about Kato varieties in higher dimensions. We shall then study the case of a specific regular quadratic polynomial automorphism $f : \mathbb{C}^3 \to \mathbb{C}^3$ in normal form with respect to [30], namely

$$f(x, y, z) = (x^2 + cy^2 + z, y^2 + x, y),$$

where $c \in \mathbb{C}$. The example we chose is essentially the only example of regular polynomial automorphism in $\mathbb{C}^3$ of degree 2. Indeed by direct computation one gets that $I^+ = [0 : 0 : 1 : 0]$ is a point, while $I^- = \{z = t = 0\}$ is a line at infinity.

In this thesis we shall construct a Kato variety associated to f^{-1} and its basin of attraction to $[0 : 0 : 1 : 0]$.

This polynomial automorphism was already considered by Oeljeklaus and Renaud in 2006 (see [52]), who constructed a different kind of 3-fold (called of Class L, see [41]) associated to the basin of attraction of f to I^-.

We shall then prove the following properties.

Theorem. *Let $f : \mathbb{C}^3 \to \mathbb{C}^3$ be the regular polynomial automorphism given by*

$$f(x, y, z) = (x^2 + cy^2 + z, y^2 + x, y),$$

and let $f_0 = F_p^{-1}$ *be the germ associated to the fixed point at infinity* $p = [0:0:1:0]$ *for* f^{-1}*. Then*

- *there exists a resolution* $f_0 = \pi \circ \sigma$, *with* π *the composition of* 6 *blow-ups (*4 *points,* 2 *curves).*

Let X *be the Kato variety associated to* (π, σ)*. Then* X *is a compactification of a fundamental domain* V *of the basin of attraction of* f_0 *to* 0*.*

Furthermore, the following properties hold for any variety Y *that is birationally equivalent to* X*:*

- Y *is a compactification of* V *(up to modifications over points in* V*), obtained by adding a suitable divisor* E*.*
- *The Kodaira dimension is* $\operatorname{kod}(Y) = -\infty$*.*
- *The first Betti number is* $b_1(Y) = b_5(Y) = 1$*.*
- E *cannot be contracted to a point.*
- *There exists a codimension* 1 *foliation.*
- *There do not exist (formal) curves in* Y *outside* E*.*

Finally, the following properties hold for X *(but they are not necessarily birationally invariant).*

- X *is obtained from* V *by adding* 6 *rational irreducible surfaces.*
- *For the Betti numbers we have* $b_2(X) = b_4(X) = 6$ *and* $b_3(X) = 0$*.*
- X *admits a global spherical shell.*

In particular we notice the main difference with the 2-dimensional case: there does not exist a contraction to a point, that is strictly related to the non-existence of a canonical "minimal model"; the other properties are however similar to the case of Kato surfaces.

Rigid germs

We presented several reasons explaining why rigid and strict germs are interesting. It is then natural to try and extend Favre's results on the classification of attracting rigid germs to higher dimensions.

A classical technique for the study of the classification of attracting germs was given by Poincaré (1883) and Dulac (1904). Roughly speaking, one can look for a general conjugation map $\Phi : (\mathbb{C}^d, 0) \to (\mathbb{C}^d, 0)$ between two germs $f, \tilde{f} : (\mathbb{C}^d, 0) \to (\mathbb{C}^d, 0)$, and study the conjugacy relation $\Phi \circ f = \tilde{f} \circ \Phi$ computing the coefficients of the formal power series involved. This approach was studied firstly for attracting invertible germs in the linearization problem, *i.e.*, one looks for a conjugation map Φ between a given germ f and its linear part $\tilde{f}(x) = df_0 x$, but it can be used also for non-invertible germs as well. It turns out that there are formal obstructions for the linearization of a germ: *resonances*.

To give definitions and state results for this subject, we need to set a few notations. Let $x = (x_1, \dots, x_d)^T$ be some coordinates in $(\mathbb{C}^d, 0)$, where ^{T} denotes the transposition of a vector or matrix. If $n = (n^1, \dots, n^d) \in \mathbb{N}^d$, we shall put $x^n = x_1^{n^1} \cdot \ldots \cdot x_d^{n^d}$, while if $N \in \mathcal{M}(c \times d, \mathbb{N})$ is a matrix, we shall denote

$$x^N = \left(x^{n_1}, \dots, x^{n_c}\right)^T,$$

where n_k is the k-st row of N. Finally, if we have two (vertical) vectors $x = (x_1, \dots, x_d)^T$ and $y = (y_1, \dots, y_d)^T$ of the same dimension d, we shall denote by $xy := (x_1 y_1, \dots, x_d y_d)^T$ the product component by component.

Definition. Let $f : (\mathbb{C}^d, 0) \to (\mathbb{C}^d, 0)$ be an attracting germ, written in suitable coordinates $x = (x_1, \dots, x_d)^T$ as

$$f(x) = \lambda x + g(x),$$

where $\lambda = (\lambda_1, \dots, \lambda_d)^T$ is the vector of eigenvalues of df_0, and $g = (g_1, \dots, g_d)^T$ includes the nilpotent linear part. Then a monomial x^n is called *resonant* for the k-th coordinate if either it arises in the k-th column of (the nilpotent part of) the Jordan form of df_0, or

$$\lambda^n = \lambda_k.$$

where $n = (n^1, \dots, n^d)$ is such that $n^1 + \dots + n^d \geq 2$.

Then the main result of Poincaré-Dulac theory is the following theorem (see [59], [54] or [6, Chapter 4]).

Theorem (Poincaré-Dulac). *Let $f : (\mathbb{C}^d, 0) \to (\mathbb{C}^d, 0)$ be an attracting germ and set $\lambda = (\lambda_1, \dots, \lambda_d)^T$ the vector of eigenvalues of df_0. Then f is holomorphically conjugated to a map $\widetilde{f} : (\mathbb{C}^d, 0) \to (\mathbb{C}^d, 0)$ of the form*

$$\widetilde{f}(x) = \lambda x + \widetilde{g}(x),$$

where $\widetilde{g} = (g_1, \dots, g_d)^T$ contains only resonant monomials.

Germs as in the Poincaré-Dulac theorem, where only resonant monomials appear, are said in *Poincaré-Dulac normal form*.

Notice that where $\lambda_k \neq 0$ the k-th coordinate of a Poincaré-Dulac normal form is polynomial; on the other hand if $\lambda_k = 0$ this theorem gives no informations on the k-th coordinate, since in this case every monomial is resonant.

It is possible however to further simplify such germs, with a "renormalization" process, *i.e.*, performing suitable changes of coordinates that preserve the property of being in Poincaré-Dulac normal form (see for example [4] for a renormalization process in the tangent-to-the-identity case, or [3] for a general procedure for formal renormalizations).

Then it is natural to study Poincaré-Dulac renormalization phenomena for attracting rigid germs. The new formal obstructions that arise are *topological resonances*.

While classic resonances for a germ f arise as algebraic relations between the eigenvalues of its linear part df_0, topological resonances appear as an algebraic relation between the non-zero eigenvalues of df_0, and the eigenvalues of a suitable matrix, called *principal part*, associated to the topological behavior of the dynamics of f.

Indeed, one of the main invariants for attracting rigid germs is the following. Given a germ $f : (\mathbb{C}^d, 0) \to (\mathbb{C}^d, 0)$, it induces an action f_* on the first fundamental group $\pi_1(\Delta^d \setminus \mathcal{C}(f^\infty))$, where Δ^d is a small open polydisc centered in 0, and $\mathcal{C}(f^\infty)$ denotes the generalized critical set. If q is the number of irreducible components of $\mathcal{C}(f^\infty)$ (thus f is said *q-reducible*), then $\pi_1(\Delta^d \setminus \mathcal{C}(f^\infty)) \cong \mathbb{Z}^q$, and f_* is represented by a suitable matrix $A \in \mathcal{M}(q \times q, \mathbb{N})$. We shall call A the *internal action* of f, and we shall say that f has *invertible* internal action if A is invertible in $\mathcal{M}(q \times q, \mathbb{Q})$. Moreover, up to permuting coordinates, we can suppose that

$$A = \left(\begin{array}{c|c} I_r & 0 \\ \hline C & D \end{array} \right),$$

for a suitable $r \leq q$. The matrix $D \in \mathcal{M}(p \times p, \mathbb{N})$ is called *principal part* of f, and $p = q - r$ its *principal rank*.

Now we can define the new resonance relation that arises in this case.

Definition. Let $f : (\mathbb{C}^d, 0) \to (\mathbb{C}^d, 0)$ be an attracting q-reducible rigid germ with invertible internal action. Let $\lambda = (\lambda_1, \dots, \lambda_s)^T$ be the vector of non-zero eigenvalues of df_0, and $D \in \mathcal{M}(p \times p, \mathbb{N})$ the principal part of f. Then a monomial x^n with $n = (n^1, \dots, n^s)$ and $x = (x_1, \dots, x_s)^T$ is called *topologically resonant* for f if λ^n is an eigenvalue for D.

In this case, as well as for classic resonances, we get a finite number of topologically resonant monomials.

Moreover, a holomorphic conjugacy result still holds.

Theorem. *Let $f : (\mathbb{C}^d, 0) \to (\mathbb{C}^d, 0)$ be an attracting rigid germ with invertible internal action of principal rank p, and set $\lambda = (\lambda_1, \dots, \lambda_s)^T$ the vector of non-zero eigenvalues of df_0. Then f is holomorphically* *conjugated to a map* $\widetilde{f} : (\mathbb{C}^d, 0) \to (\mathbb{C}^d, 0)$ *of the form*

$$\widetilde{f}(x) = \Big(\lambda x_{\leq s} + \delta(x_{\leq s}), \alpha x^B_{\leq s+p}\big(1 + \widetilde{g}(x_{\leq s})\big), \widetilde{h}(x)\Big)^T,$$

where $x = (x_1, \ldots, x_d)^T$, $x_{\leq k} := (x_1, \ldots, x_k)^T$, *and* $\delta = (\delta_1, \ldots, \delta_s)^T$ *has only resonant monomials.*

Moreover $\alpha = (\alpha_{s+1}, \ldots, \alpha_{s+p})^T \in (\mathbb{C}^*)^p$, $\widetilde{g} = (g_{s+1}, \ldots, g_{s+p})^T$ *where* g_k *has only topologically resonant monomials for every* k, *and* $\widetilde{h} = (h_{s+p+1}, \ldots, h_d)^T$. *Finally,*

$$B = \big(\ C' \mid D\ \big),$$

where D *denotes the principal part of* f.

The main feature for this normal form is that now $\widetilde{g}$ depends only on the first s coordinates. In particular, the coordinates f_k of f with $k = s+1, \ldots, s+p$ depends only on the first $s+p$ coordinates. This allows for example to construct in a suitable neighborhood of $0 \in \mathbb{C}^d$ an f-invariant foliation of codimension $s+p$, induced by $dx_1 \wedge \ldots \wedge dx_{s+p}$, that is actually a subfoliation of the f-invariant foliation of codimension s induced by $dx_1 \wedge \ldots \wedge dx_s$.

As a consequence of this result, we shall give the (almost) complete holomorphic classification of q-attracting rigid germs in $\mathbb{C}^d$ with invertible internal action, for $q = d-1$ and $q = d$.

In particular, for $d = 3$ we get the classification for 2-reducible and 3-reducible attracting rigid germs with invertible internal action, while for 0-reducible, *i.e.*, invertible germs, the classification follows from the Poincaré-Dulac theorem.

This work is subdivided into 4 chapters. In the first one, containing just some background material, we recall some notations and standard results on holomorphic dynamics, algebraic geometry, algebraic topology, complex manifolds. The second chapter is devoted to valuative tree theory, and the study of semi-superattracting germs in $\mathbb{C}^2$. The third chapter deals with the classification of holomorphic rigid germs in $\mathbb{C}^n$, while the last chapter is dedicated to Kato varieties.

Acknowledgements

The material presented in this thesis comes from the research I pursued during my doctoral years at the Scuola Normale Superiore di Pisa. My gratitude goes to many people, who in different ways helped me in reaching this goal.

First of all, I would like to particularly thank my advisor Marco Abate, who supported me during my undergraduate and PhD studies, shared a lot of enlightening discussions, and guided me patiently during the drafting of this thesis.

The last two chapters of this thesis were developed during the last year, while visiting the École Normale Supérieure de Paris, under the supervision of Charles Favre. I owe him my deepest gratitude for the ideas he shared with me, the geometrical insight and all the time he devoted to me.

I am also grateful to Mattias Jonsson, for several useful conversations and his warm hospitality during my visiting period at the University of Michigan.

I am definitely indebted to other professors and colleagues, who helped me with fruitful discussions and insights, and I thank them all.

I owe my gratitude to Scuola Normale Superiore for its support during these beautiful years.

A special thanks goes to my family, who supported and trusted me for all these years, and all my friends spread all over the world, who shared many interests and unforgivable days with me.

Chapter 1
Background

1.1. Holomorphic dynamics

1.1.1. Local holomorphic dynamics

In this section we shall recall a few definitions and known results in local discrete holomorphic dynamics.

Local discrete holomorphic dynamics deals with holomorphic maps $f : \mathbb{C}^d \to \mathbb{C}^d$ (with $d \geq 1$ the dimension) with a fixed point at the origin $0 \in \mathbb{C}^d$. We are interested into orbits of f, *i.e.*, the behavior of the iterates $f^{\circ n}$, in a neighborhood of the fixed point, and hence we consider a germ $f : (\mathbb{C}^d, 0) \to (\mathbb{C}^d, 0)$. Usually two such germs are considered to be equivalent if one can get one from another through a "change of coordinates".

Definition 1.1.1. Let $f, g : (\mathbb{C}^d, 0) \to (\mathbb{C}^d, 0)$ be two holomorphic germs. We shall say that f and g are (**holomorphically**, **topologically**, **formally**) **conjugated** if there exists a (biholomorphism, homeomorphism, formal invertible map) $\phi : (\mathbb{C}^d, 0) \to (\mathbb{C}^d, 0)$ such that

$$\phi \circ f = g \circ \phi.$$

Then one of the main goals of local holomorphic dynamics is to find "normal forms", *i.e.*, a (possibly small) family of germs called normal forms, such that a general germ can be conjugated to one of these.

It is not always possible to find explicit normal forms, but still there are some objects, called "invariants", very useful to distinguish two germs that are not conjugated.

Definition 1.1.2. Let $\mathrm{End}(\mathbb{C}^d, 0)$ be the set of holomorphic germs of the form $f : (\mathbb{C}^d, 0) \to (\mathbb{C}^d, 0)$. A map $I : \mathrm{End}(\mathbb{C}^d, 0) \to C$, with C a suitable set, is called (**holomorphic**, **topological**, **formal**) **invariant** if $I(f) = I(g)$ whenever f and g are (holomorphically, topologically, formally) conjugated.

A set of invariants I is called **complete** if also the converse holds: two germs f and g are conjugated if and only if $I(f) = I(g)$.

Example 1.1.3. Let $f : (\mathbb{C}^d, 0) \to (\mathbb{C}^d, 0)$ be a holomorphic germ. Then the Jordan form of the differential df_0 of f in 0 is a holomorphic and formal (but not topological) invariant.

Definition 1.1.4. Let $f : (\mathbb{C}^d, 0) \to (\mathbb{C}^d, 0)$ be a holomorphic germ. If f is holomorphically (resp., formally) conjugated to its linear part, we shall say that f is holomorphically (resp., formally) **linearizable**.

1.1.2. Local dynamics in one complex variable

We shall give here a short exposition of classical results in local holomorphic dynamics in one complex variable. For a complete exposition on this field, we refer to [51], [12] or [7].

Definition 1.1.5. Let $f : (\mathbb{C}, 0) \to (\mathbb{C}, 0)$ be a holomorphic germ with a fixed point $0 \in \mathbb{C}$. We shall call **multiplier** the value $\lambda = f'(0)$. In particular, such a germ can be written as a power series

$$f(z) = \lambda z + a_2 z^2 + a_3 z^3 + \dots.$$

As we have seen in Example 1.1.3, the multiplier of a germ is a holomorphic and formal conjugacy invariant. Depending on the multiplier, we have the following subdivision of the set of germs in a few classes.

Definition 1.1.6. Let $f : (\mathbb{C}, 0) \to (\mathbb{C}, 0)$ be a holomorphic germ with multiplier $\lambda \in \mathbb{C}$. Then f is called

- **attracting** if $|\lambda| < 1$, and in particular it is called **superattracting** if $\lambda = 0$;
- **repelling** if $|\lambda| > 1$;
- **rational neutral** or **parabolic** if λ is a root of unity, and in particular it is called **tangent to the identity** if $\lambda = 1$;
- **irrational neutral** or **elliptic** if $|\lambda| = 1$ but λ is not a root of unity.

The first result in (local) holomorphic dynamics was given then by Kœnigs in 1884, and it is the classification of germs $f : (\mathbb{C}, 0) \to (\mathbb{C}, 0)$ in the (invertible) attracting and repelling case (see [44] for the original paper, and [51, Theorem 8.2] for a modern exposition).

Theorem 1.1.7 (Kœnigs). *Let $f : (\mathbb{C}, 0) \to (\mathbb{C}, 0)$ be a holomorphic germ such that the multiplier $\lambda := f'(0)$ is such that $|\lambda| \neq 0, 1$. Then f is holomorphically conjugated to the linear part $z \mapsto \lambda z$.*

Hence, every germ in the hypothesis of Theorem 1.1.7 (Kœnigs) is holomorphically (and hence formally) linearizable.

Then in 1904, Böttcher solved the superattracting case. See [10] for the original paper, and [51, Theorem 9.1] for a modern exposition of the proof.

Theorem 1.1.8 (Böttcher). *Let $f : (\mathbb{C}, 0) \to (\mathbb{C}, 0)$ be a holomorphic germ of the form*

$$f(z) = a_p z^p + a_{p+1} z^{p+1} + \dots,$$

with $p \geq 2$ and $a_p \neq 0$. Then f is holomorphically conjugated to the map $z \mapsto z^p$.

The neutral case is trickier and probably more fascinating.

In the rational case, the first easy result is the following (see for example [50, Proposition 1.6]).

Proposition 1.1.9. *Let $f : (\mathbb{C}, 0) \to (\mathbb{C}, 0)$ a parabolic germ. Then f is holomorphically linearizable if and only if it is topologically linearizable, if and only if there exists n such that $f^{\circ n}$ is a linear map.*

This result already shows how the situation is not so trivial in the parabolic case as it was in the attracting/repelling case, since a parabolic germ is almost never linearizable.

From now on we shall focus on the tangent to the identity case: every parabolic germ can be considered as a tangent to the identity germ up to taking a suitable iterate.

Thanks to Proposition 1.1.9, a tangent to the identity germ f is (holomorphic, topologically) linearizable if and only if f is the identity map.

Definition 1.1.10. Let $f : (\mathbb{C}, 0) \to (\mathbb{C}, 0)$ be a tangent to the identity (holomorphic) germ, written on the form:

$$f(z) = z(1 + a_k z^k + a_{k+1} z^{k+1} + \dots),$$

with $k \geq 1$ and $a_k \neq 0$. Then $k + 1$ is called the **parabolic multiplicity** of f.

The parabolic multiplicity of a germ is actually a holomorphic, formal and topological invariant, and it is a complete invariant for the topological classification.

For the formal classification, it turns out that there exists another invariant, called index.

Definition 1.1.11. Let $f : (\mathbb{C}, 0) \to (\mathbb{C}, 0)$ be a tangent to the identity (holomorphic) germ.

We shall call **index** of f the complex value

$$\beta := \frac{1}{2\pi i} \int_\gamma \frac{dz}{z - f(z)},$$

with γ a positive loop around 0.

Definition 1.1.12. A **parabolic domain** in $\mathbb{C}$ is a simply connected open domain Δ such that $0 \in \partial\Delta$.

A parabolic domain Δ is said to be an **attracting petal** (resp., **repelling petal**) for a tangent to the identity map f if $f(\Delta) \subset \Delta$ and $f^{\circ n}(x) \to 0$ for every $x \in \Delta$ (resp., the same for f^{-1}).

For the dynamics of such germs, the main result is given by the following theorem (see [46, 47]) and Fatou (see [24] for the original papers, and [51, Theorems 10.7 and 10.9] for a modern exposition).

Theorem 1.1.13 (Leau, Fatou). *Let $f : (\mathbb{C}, 0) \to (\mathbb{C}, 0)$ be a tangent to the identity germ with parabolic multiplicity $k + 1$. Then there exists k attracting petals, and k repelling petals, such that in every petal f is holomorphically conjugated to the map*

$$z \mapsto z + 1.$$

Distinct attracting petals are disjoint one to another, and the same holds for repelling petals. Attracting and repelling petals form a punctured neighborhood of 0.

The holomorphic classification of tangent to the identity germs is very complex, and depends on a third invariant (besides the parabolic multiplicity and the index), called **sectorial invariant**, and introduced independently by Écalle and Voronin, with different techniques. As a matter of fact, Écalle developed the theory of "resurgent functions" (see [21], [22], and [23]), while Voronin used a more analytic approach, studying Stokes phenomena (see [61]).

Remark 1.1.14. Theorem 1.1.13 (Leau, Fatou) gives us k attracting petals P_{2j} and k repelling petals P_{2j+1} with $j = 0, \dots, k-1$, that alternate to cover a punctured neighborhood of the origin; moreover it gives $2k$ conjugations $\phi_j : P_j \to \mathbb{C}$ for $j = 0, \dots, 2k-1$ between f and the translation map $z \mapsto z+1$ when j is even, and $z \mapsto z-1$ when j is odd. In particular one can consider the transition maps in the intersection of two adjacent petals, $\phi_{j+1}^{-1} \circ \phi_j$ (where we set $\phi_0 = \phi_{2k}$). Roughly speaking, the sectorial invariant describes the analytic behavior of the family of these transition maps. Its moduli space is infinite-dimensional, and the sectorial invariant is not easy to compute. For these reasons the research in this field is still open, trying to find coarser but computable invariants for these germs.

We will not deal with the neutral irrational case, that has its peculiarities; in particular the holomorphic classification involves number theory, diophantine conditions and other fields of mathematics. Just to give an idea of how these aspects can merge together, if we set $\lambda = e^{2\pi i\alpha}$ with $\alpha \in \mathbb{R} \setminus \mathbb{Q}$ the multiplier of a germ f, then the dynamical behavior (linearization, existence of periodic orbits near the fixed point, etc.) of f is strictly related to how α can be well approximated by rational numbers, and this gives some diophantine conditions, or even more complex conditions (for example the *Brjuno condition* or the *Perez-Marco condition*) on α. We refer to [50] for an exposition of this matter.

In the following we shall need the formal classification of germs in dimension 1, so we shall state it here, also for summing up all the cases we just described.

Proposition 1.1.15 (Formal classification in $(\mathbb{C}, 0)$). *Let $f : (\mathbb{C}, 0) \to (\mathbb{C}, 0)$ be a holomorphic germ, and denote by $\lambda = f'(0)$ the multiplier. Then*

(i) *if $\lambda = 0$, then f is formally conjugated to $x \mapsto x^p$ for a suitable $p \geq 2$;*

(ii) *if $\lambda \neq 0$, and $\lambda^r \neq 1$ for any $r \in \mathbb{N}^*$, then f is formally conjugated to $x \mapsto \lambda x$;*

(iii) *if $\lambda^r = 1$, then there exist (unique) $s \in r\mathbb{N}^*$ and $\beta \in \mathbb{C}$ such that f is formally conjugated to $x \mapsto \lambda x(1 + x^s + \beta x^{2s})$.*

1.1.3. Local dynamics in several complex variables

In this subsection we shall fix some notations on holomorphic germs in higher dimensions.

First of all, unless otherwise specified, we shall consider only dominant holomorphic germs.

Definition 1.1.16. Let $f : (\mathbb{C}^2, 0) \to (\mathbb{C}^2, 0)$ be a holomorphic germ. Then f is **dominant** if $\det(df_p)$ is not identically zero.

As in the one dimensional case with the multiplier, we are interested into subdividing holomorphic germs depending on (the eigenvalues, the Jordan form of) the differential at 0 (see Example 1.1.3).

Definition 1.1.17. Let $f : (\mathbb{C}^d, 0) \to (\mathbb{C}^d, 0)$ be a holomorphic germ, and let us denote by $\mathrm{Spec}(df_0) = \{\lambda_1, \dots, \lambda_d\}$ the set of eigenvalues of df_0. Then f is said:

- **attracting** if $|\lambda_i| < 1$ for $i = 1, \dots, d$;
- **superattracting** if $df_0 = 0$;
- **nilpotent** if df_0 is nilpotent (*i.e.*, $df_0^d = 0$; in particular, superattracting germs are nilpotent germs);
- **tangent to the identity** if $df_0 = \mathrm{Id}$;
- **hyperbolic** if $|\lambda_i| \neq 1$ for every i; it is said a **saddle**, or **Siegel hyperbolic** if it is hyperbolic and there is at least one eigenvalue with modulus smaller than 1, and another with modulus greater than 1.

We shall also use the following definition in the 2-dimensional case.

Definition 1.1.18. Let $f : (\mathbb{C}^2, 0) \to (\mathbb{C}^2, 0)$ be a holomorphic germ. Then f is said:

- **semi-superattracting** if $\mathrm{Spec}(df_0) = \{0, \lambda\}$, with $\lambda \neq 0$;
- **of type** $(0, D)$ if $\mathrm{Spec}(df_0) = \{0, \lambda\}$ and $\lambda \in D$, with $D \subset \mathbb{C}$ a subset of the complex plane.

In particular the semi-superattracting germs are the ones of type $(0, \mathbb{C}^*)$.

Studying the dynamics of an attracting germ $f : (\mathbb{C}^d, 0) \to (\mathbb{C}^d, 0)$, it is natural to look at its basin of attraction U, and fundamental domains of U.

Definition 1.1.19. Let $f : (\mathbb{C}^d, 0) \to (\mathbb{C}^d, 0)$ be an attracting holomorphic germ (with $f(0) = 0$), and B a ball in $\mathbb{C}^d$ where f is defined. We shall call the set

$$U = U_f(0) := \{x \in B \mid f^{\circ n} \to 0\}$$

the **basin of attraction** of f to 0.

If we take the quotient of U by the action of f, we get the **space of orbits** $U/\rangle f \langle$ of U. Then a **fundamental domain** of U is a open dense subset of the space of orbits of U.

The study of holomorphic germs, even in dimension 2, is far to be complete, and only a few classes have been classified, such has the attracting invertible germs. Another class of holomorphic germs that has been classified (see [25]) is the one of attracting rigid germs (that contains all attracting invertible germs).

Definition 1.1.20. Let $f : (\mathbb{C}^2, 0) \to (\mathbb{C}^2, 0)$ be a (dominant) holomorphic germ. We denote by $\mathcal{C}(f) = \{z \mid \det(df_z) = 0\}$ the **critical set** of f, and by $\mathcal{C}^\infty(f) = \bigcup_{n\in\mathbb{N}} f^{-n}\mathcal{C}(f)$ the **generalized critical set** of f. Then a (dominant) holomorphic germ f is **rigid** if:

(i) $\mathcal{C}^\infty(f)$ (is empty or) has simple normal crossings (**SNC**) at the origin; and
(ii) $\mathcal{C}^\infty(f)$ is forward f-invariant.

Remark 1.1.21. In [25], the condition (ii) is not explicitly stated in the definition of a rigid germ, but it is implicitly used. The second property does not follow from the first one: if for example we consider the map $f(z, w) = (\lambda z^p, z(1 + w^2))$, with $p \geq 1$ and $\lambda \in \mathbb{C}^*$, the generalized critical set is $\{zw = 0\}$, but $f(z, 0) = (\lambda z^p, z)$ and hence $\mathcal{C}^\infty(f)$ is not forward f-invariant.

As anticipated in the introduction, rigid germs shall be the main object of this thesis. We shall deal more deeply with (attracting) rigid germs in higher dimensions in Chapter 3.

1.1.4. Parabolic curves

Studying the tangent to the identity case in dimension $d \geq 2$, one would like to find "petals" such as in dimension one. Parabolic domains and petals are objects of dimension one, but they can be clearly generalized in higher dimensions.

Definition 1.1.22. A **parabolic domain** in $\mathbb{C}^d$ is a simply connected open domain $\Delta \subset \mathbb{C}^d$ such that $0 \in \partial\Delta$.

We shall call d the **dimension** of the parabolic domain.

A parabolic domain Δ is said to be **attracting** for a holomorphic map (germ) $f : (\mathbb{C}^d, 0) \to (\mathbb{C}^d, 0)$ if $f(\Delta) \subset \Delta$ and $f^{\circ n}(x) \to 0$ for every $x \in \Delta$.

Sometimes it is not possible to find attracting parabolic domains for a (tangent to the identity) germ f of maximal dimension, but we can try to find a parabolic domain of dimension $k \leq d$ embedded in our space of dimension d.

Definition 1.1.23. A **parabolic k-variety** for a (tangent to the identity) germ $f : (\mathbb{C}^d, 0) \to (\mathbb{C}^d, 0)$ at the origin is a injective holomorphic map $\varphi : \Delta \to \mathbb{C}^d$ satisfying the following properties:

(i) Δ is a parabolic domain of dimension k;
(ii) φ is continuous at the origin, and $\varphi(0) = 0$;
(iii) $\varphi(\Delta)$ is invariant under f, and $f^{\circ n}(z) \to 0$ for every $z \in \varphi(\Delta)$.

We shall call **parabolic curves** the parabolic 1-varieties.

1.1.5. Stable and unstable manifolds

In this subsection, we shall present a classical result in holomorphic dynamics, the *Stable/Unstable manifold theorem*, stated for holomorphic maps, and its generalization, the *Hadamard-Perron theorem*. For references, see [1].

First of all, we recall what a dynamical system is.

Definition 1.1.24. Let $f : X \to X$ a continuous map in a topological space X. We shall call the pair (X, f) a (continuous, discrete) **dynamical system**.

Definition 1.1.25. Let (X, f) a dynamical system. then a subset $U \subseteq X$ of X is called **forward invariant** if $f(U) \subseteq U$, **backward invariant** if $f^{-1}(U) \subseteq U$.

We shall define first the Stable set of a dynamical system (X, f).

Definition 1.1.26. Let (X, f) a dynamical system in a metric space X that fixes a point p, and denote by $\|x\|$ the distance between x and p.

Then we shall call the set

$$W_f^s := \left\{ x \in X \mid \lim_{n \to +\infty} \left\| f^n(x) \right\| = 0 \right\}$$

the **stable set** for f (at p).

Defining the Unstable set is straightforward when f is invertible.

Definition 1.1.27. Let (X, f) a dynamical system in a metric space X where f is a homeomorphism that fixes a point p.

Then we shall call the set

$$W_f^u := W_{f^{-1}}^s$$

the **unstable set** for f (at p).

For defining the unstable set in the general case, we need a few definitions.

Definition 1.1.28. Let (X, f) be a dynamical system. The **dynamical completion** of (X, f) is the dynamical system $(\hat{X}, \hat{f})$, where

$$\hat{X} = \left\{ \hat{x} = (x_n)_{n \in \mathbb{Z}} \mid X \ni x_{n+1} = f(x_n) \text{ for all } n \in \mathbb{Z} \right\},$$

and $\hat{f}$ is just the left shift $\hat{f}((x_n)) := (f(x_n)) = (x_{n+1})$.

We shall call a point $\hat{x} \in \hat{X}$ a **history** (of x_0), and we shall call **canonical projection** the map $\pi : \hat{X} \to X$ defined by $\pi(\hat{x}) = x_0$.

Remark 1.1.29. If (X, f) is a dynamical system, and $(\hat{X}, \hat{f})$ is its dynamical completion, then we have that $\hat{f}$ is a homeomorphism of $\hat{X}$. If moreover X is endowed with a distance d, then $\hat{X}$ inherits a distance function too. For example, if d is bounded (that can be always be assumed, up to taking a distance that induces the same topology), we can define

$$d(\hat{x}, \hat{y}) := \sum_{n \in \mathbb{Z}} a_n d(x_n, y_n),$$

where $\sum_{n \in \mathbb{Z}} a_n < +\infty$.

Finally, if p is a fixed point for $f : X \to X$, then the history $\hat{p} = (p)_{n \in \mathbb{Z}}$ is a fixed point for $\hat{f} : \hat{X} \to \hat{X}$.

Definition 1.1.30. Let (X, f) be a dynamical system in a metric space X that fixes a point p, and $(\hat{X}, \hat{f})$ its dynamical completion.

Then we shall call the set

$$W^u_f := \pi(W^u_{\hat{f}})$$

the **unstable set** for f (at p), where $W^u_{\hat{f}}$ denotes the unstable set for $\hat{f}$ at $\hat{p} = (p)_{n \in \mathbb{Z}}$, and $\pi : \hat{X} \to X$ the canonical projection.

We shall obtain the Stable/Unstable manifold theorem as a corollary of the Hadamard-Perron theorem, but to state the latter, we need another definition.

Definition 1.1.31. Given $0 < \lambda < \mu$, a linear map $L : V \to V$ of a normed vector space (over $\mathbb{C}$) admits a (λ, μ)**-splitting** if there is a decomposition $V = E^+ \oplus E^-$ such that $L(E^+) = E^+$, $L(E^-) \subseteq E^-$, $\|L(v)\| \leq \lambda \|v\|$ for all $v \in E^-$ and $\|L(v)\| \geq \mu \|v\|$ for all $v \in E^+$.

The dimensions $\dim E^-$ and $\dim E^+$ are called the **stable** and **unstable dimension** of the splitting.

Theorem 1.1.32 (Hadamard-Perron, see, *e.g.*, [1, Theorem 3.1.4]). *Let $f : (\mathbb{C}^d, 0) \to (\mathbb{C}^d, 0)$ be a holomorphic germ, and $0 < \lambda < \mu$ such that df_0 admits a (λ, μ)-splitting $\mathbb{C}^d = E^- \oplus E^+$ of unstable dimension k.*

Then, up to replacing λ and μ with $\lambda(1+\varepsilon) < \mu(1-\delta)$, with $\varepsilon, \delta > 0$ arbitrarily small, there exists (locally at 0) a unique k-dimensional C^1 manifold W^+ and a unique $(d-k)$-dimensional C^1 manifold W^- such that:

(i) $0 \in W^+ \cap W^-$;
(ii) *$W^\pm$ is tangent to $E^\pm$ at the origin;*
(iii) *$f(W^-) \subseteq W^-$, $f(W^+) = W^+$ and $f|_{W^+}$ is invertible;*

(iv) $\|f(z)\| \leq \lambda \|z\|$ *for all* $z \in W^-$, *and* $\|f(z)\| \geq \mu \|z\|$ *for all* $z \in W^+$;

(v) *we have that* $z_0 \in W^-$ *if and only if there are* $\lambda \leq \nu < \mu$ *and* $C > 0$ *such that* $\|f^{\circ n}(z_0)\| \leq C\nu^n \|z_0\|$ *for all* $n \geq 1$;

(vi) *we have that* $z_0 \in W^+$ *if and only if there are* $\lambda < \nu \leq \mu$, $C > 0$ *and an history* $\hat{z}$ *for* z_0 *such that* $\|z_{-n}\| \leq C\nu^{-n} \|z_0\|$ *for all* $n \geq 1$;

(vii) *the manifolds* $W^\pm$ *are of class* C^∞;

(viii) *fix a neighborhood* U *of the origin where* f *is holomorphic; then the manifold* W^+ *(resp.,* W^-*) is a complex submanifold at all the points* z *such that there exists an history* $\hat{z}$ *of* z *such that* $p_m \in U$ *for all* $m \leq 0$ *(resp., for all* $m \geq 0$*).*

Theorem 1.1.33. (Stable-Unstable manifold Theorem, [1, Theorem 3.1.2]. *Let* $f : (\mathbb{C}^d, 0) \to (\mathbb{C}^d, 0)$ *be a holomorphic germ, and* $0 < \lambda < 1 < \mu$ *such that* df_0 *admits a* (λ, μ)*-splitting* $\mathbb{C}^d = E^- \oplus E^+$ *of unstable dimension* k.

Then the stable set W^s *and the unstable set* W^u *are (resp.,* $(d-k)$*-dimensional and* k*-dimensional) complex manifolds forward (resp., backward) invariant for* f.

1.2. Algebraic geometry

1.2.1. Divisors and line bundles

In this section we just set some notations. For a basic exposition of this subject, see [56, Chapter III, Section 1] and [31, Chapter 1, Section 1].

Definition 1.2.1. Unless otherwise stated, we shall mean by n**-variety** a compact complex (possibly singular) manifold of complex dimension n, and by n**-manifold** a smooth n-variety. Since we shall always consider divisors in non-singular varieties, with **divisor** we shall mean either a Weil or Cartier divisor. If D is a divisor in a manifold X, we shall denote by $[D]$ the (complex) line bundle associated to D.

We shall use additive notations for line bundles: if L_1 and L_2 are two line bundles over X, we shall denote $L_1 + L_2 := L_1 \otimes L_2$ their tensor product.

Hence with our notations, if L is a line bundle, than its dual $L^* = -L$, while if $L_i = [D_i]$ for $i = 1, 2$, then $L_1 + L_2 = [D_1 + D_2]$.

We shall usually consider divisors up to linear equivalence, and line bundles up to isomorphism. We shall denote by $\mathrm{Pic}(X)$ the **Picard group** of X, *i.e.*, the group (with respect to the tensor product) of holomorphic line bundles over X up to isomorphism.

We shall also use the following property of divisors.

Proposition 1.2.2. *Let X be a complex variety, D a divisor in X, and $V \subset X$ a hypersurface that intersects transversely the support of D. Then*

$$[D]|_V = [D \cap V].$$

1.2.2. Blow-ups and Modifications

Definition 1.2.3. Let X be a complex manifold of (complex) dimension $n \geq 2$, and let $Y \subset X$ be a complex submanifold of dimension $0 \leq k \leq n-2$. We denote by $E := \mathbb{P}(\mathcal{N}_{Y \subset X})$ the projectivization of the normal bundle of Y in X. Then $E \to Y$ is a bundle over Y with fiber $\mathbb{P}^{n-k-1}$. We shall denote by $E_p := \mathbb{P}(T_pX/T_pY)$ the fiber of $E \to Y$ in a point p.

We can equip $\widetilde{X} := X \setminus Y \cup E$ with a complex structure. Every chart (U, ϕ) for X that does not intersect Y will be a chart for $\widetilde{X}$ too. If (U, ϕ) is a chart of X that intersects Y, and we write $\phi = (x_1, \ldots, x_n)$, we can suppose that $\phi(Y \cap U) = \{x_{k+1} = \ldots = x_n = 0\}$. Then $E|_{Y \cap U}$ has as coordinates

$$\left(x_1, \ldots, x_k, \left[\frac{\partial}{\partial x_{k+1}} : \ldots : \frac{\partial}{\partial x_n}\right]\right).$$

We shall consider $X_j = \{x_j = 0\} \subset U$, $L_j = \mathbb{P}(\ker dx_j) = \{\partial/\partial x_j = 0\} \subset E|_{Y \cap U}$, and $U_j := (U \setminus X_j) \cup (E|_{Y \cap U} \setminus L_j)$, for $j = k+1, \ldots, n$. Then we define $\chi_j : U_j \to \mathbb{C}^n$, as

$$\chi_j(q)_h = \begin{cases} x_h(q) & \text{if } h \leq k \text{ or } h = j, \\ x_h(q)/x_j(q) & \text{if } h > k \text{ and } h \neq j, \end{cases}$$

if $q \in U \setminus X_j$; and

$$\chi_j([v])_h = \begin{cases} 0 & \text{if } h \leq k \text{ or } h = j, \\ d(x_h)_p(v)/d(x_j)_p(v) & \text{if } h > k \text{ and } h \neq j, \end{cases}$$

if $[v] \in E_p \setminus L_j$.

So $\widetilde{X}$, with the natural projection $\pi : \widetilde{X} \to X$, is called the **blow-up** of X along Y.

The manifold $\widetilde{X}$ is called the **total space** of the blow-up, the submanifold $Y \subset X$ is called the **center**, while $E = \pi^{-1}(Y) \subset \widetilde{X}$ is called the **exceptional divisor**.

Definition 1.2.4. Let X be a complex n-manifold, Y a complex submanifold of dimension k, and let $\pi : \widetilde{X} \to X$ be the blow-up of X along Y. Let moreover C be an analytic subset of X not contained into Y. The **strict transform** of C is the set $\widetilde{C} = \overline{\pi^{-1}(C \setminus Y)} \subset \pi^{-1}(C)$.

Definition 1.2.5. Let X be a complex n-manifold, and $p \in X$ a point. We call a holomorphic map $\pi : Y \to (X, p)$ a **modification** over p if π is a composition of blow-ups, with the first one being a point blow-up along p, and such that π is a biholomorphism outside $\pi^{-1}(p)$. We call $\pi^{-1}(p)$ the **exceptional divisor** of π, and we call every irreducible component of the exceptional divisor an **exceptional component**. We shall denote by Γ_π^* the set of all exceptional components of a modification π.

We shall call a point $p \in \pi^{-1}(0)$ on the exceptional divisor of a modification $\pi : X \to (\mathbb{C}^n, 0)$ an **infinitely near point** (we shall consider $0 \in \mathbb{C}^n$ as an infinitely near point too).

We shall also call **weight** of a modification π the number of blow-ups the modification is composed by, or equivalently the number of irreducible components of the exceptional divisor of π, and we shall denote it by $\mathrm{weight}(\pi)$.

Remark 1.2.6. The situation in dimension 2 is much simpler. Indeed, we can blow-up only points, and then the exceptional components are all biholomorphic to $\mathbb{P}^1$. Moreover if $\pi : Y \to (X, p)$ is a modification over a complex 2-manifold, then a point in the exceptional divisor can belong to either one or two exceptional components.

Definition 1.2.7. We shall denote by $\mathfrak{B}$ the set of all modifications over $0 \in \mathbb{C}^2$. A point $p \in E = \pi^{-1}(0)$ for a $\pi \in \mathfrak{B}$ is called **free point** if it belongs to only one exceptional component, or equivalently if it is a smooth point in E; it is called **satellite point** if it belongs to two exceptional components (that meet transversely), and hence it is a critical point in E. Satellite points are also known as **corners** in literature.

We can equip $\mathfrak{B}$ with a partial ordering.

Definition 1.2.8. Let $\pi_1, \pi_2 \in \mathfrak{B}$ be two modifications; we say that $\pi_1 \trianglerighteq \pi_2$ if $\pi_1 = \pi_2 \circ \widetilde{\pi}$, with $\widetilde{\pi}$ a composition of point blow-ups over the exceptional divisor of π_2.

This partial order could be defined also for modifications in higher dimensions, giving to them the structure of a poset. But in the 2-dimensional case, $\mathfrak{B}$ is not simply a poset, but it has better properties.

Definition 1.2.9. A poset B is a **direct system** if every (non-empty) finite subset of B admits a supremum; it is said an **inverse system** if every (non-empty) subset of B admits an infimum.

Proposition 1.2.10 ([26, Lemma 6.1]). *The set of modifications* $(\mathfrak{B}, \trianglelefteq)$ *is a direct and an inverse system.*

Definition 1.2.11. Let $\pi_1, \pi_2 \in \mathfrak{B}$ be two modifications. We shall call the supremum $\pi_1 \vee \pi_2$ the **join** between π_1 and π_2, while we shall denote their infimum by $\pi_1 \wedge \pi_2$.

Dual Graph and Dual Complex

Definition 1.2.12. Let X be a complex n-manifold, and D a SNC divisor in X. Let $D_1, \dots, D_h$ be all the irreducible hypersurfaces such that $D = \sum_{i=1}^{h} a_i D_i$. We shall call **dual complex** of D the complex Γ_D constructed as follows. The vertices of Γ_D are all the components $D_1, \dots D_h$. Moreover $D_{i_1}, \dots, D_{i_k}$ are connected by a $k-1$-simplex for each irreducible component of $D_{i_1} \cap \dots \cap D_{i_k} \neq \emptyset$.

Definition 1.2.13. Let X be a complex 2-manifold, and D a SNC divisor in X. Then the dual complex of D is called **dual graph** of D.

Definition 1.2.14. Let π be a modification over $(\mathbb{C}^n, 0)$. We shall call **dual complex** (resp., **dual graph** if $n = 2$) of π the dual complex (resp., dual graph) of its exceptional divisor $E = \pi^{-1}(0)$.

Remark 1.2.15. Let us focus on the 2-dimensional case. We can construct the dual graph of a modification recursively on the number of point blow-ups that make up the modification.

Let $\pi \in \mathfrak{B}$ be a modification (over 0). Then the first blow-up is the blow-up of the origin. This simply gives us an exceptional component E_0, and no edges. Now suppose that we have constructed the dual graph $\Gamma^*_{\pi_1}$ of a modification $\pi_1 \trianglelefteq \pi$, and analyze a point blow-up over $\pi_1^{-1}(0)$. If p is a free point, then it belongs to just an exceptional component, say E: we have to add a vertex (E_p) and an edge between E and E_p. If p is a satellite point, then it belongs to two exceptional components, say E and F: we have to add a vertex (E_p), and change the edge between E and F with two edges, one between E and E_p, and another between E_p and F.

1.2.3. Canonical and Normal Bundles

We shall just fix some notations for canonical and normal bundles.

Definition 1.2.16. Let X be a compact complex n-manifold, and $V \subset X$ a submanifold of X. Then the **normal bundle** of V in X is the bundle $\mathcal{N}_{V \subset X} := TX|_V / TV$. We shall usually use the notation $\mathcal{N}_V$ instead of $\mathcal{N}_{V \subset X}$ if the ambient space is clear from the contest.

Definition 1.2.17. Let X be a compact complex n-manifold. Then the **canonical bundle** of X is the linear bundle $\mathcal{K}_X := \wedge^n T^* X$, where $T^* X$ is the holomorphic cotangent bundle over X.

Example 1.2.18. Here are some examples for canonical bundles in easy cases.

- The canonical bundle of $\mathbb{P}^n$ is $\mathcal{K}_{\mathbb{P}^n} = -(n+1)[H]$, where H is an hyperplane in $\mathbb{P}^n$ (see [31, p.146]).

- If we consider X and Y two complex varieties, then we have $\mathcal{K}_{X\times Y} = \pi_1^*\mathcal{K}_X + \pi_2^*\mathcal{K}_Y$, where π_1 and π_2 are the projections from $X \times Y$ respectively to X and Y.

The next proposition computes the canonical class of a blown-up manifold; for proofs, see [31, p.608].

Proposition 1.2.19. *Let X be a compact complex manifold of dimension n, and $V \subset X$ a submanifold of dimension k. Let $\pi : \widetilde{X} \to X$ be the blow-up of X along V. Then for the canonical classes, we have*

$$\mathcal{K}_{\widetilde{X}} = \pi^*\mathcal{K}_X + (n-k-1)E,$$

where $E = \pi^{-1}(V)$ denotes the exceptional divisor of π.

We shall now state two classical results, about the connection between normal and canonical bundles, known as Adjunction Formulae. For proofs, see for example [31, pp. 146–147].

Proposition 1.2.20 (Adjunction formula 1**).** *Let X be a compact complex manifold, and $V \subset X$ a smooth analytic hypersurface. Then*

$$\mathcal{N}_V = [V]|_V,$$

where $\mathcal{N}_V$ is the normal bundle of V in X, and $[V]$ is the line bundle on X associated to V as a divisor of X.

Proposition 1.2.21 (Adjunction formula 2**).** *Let X be a compact complex manifold, and $V \subset X$ a smooth analytic hypersurface. Then*

$$\mathcal{K}_V = (\mathcal{K}_X \otimes [V])|_V.$$

We shall need the computation of the canonical bundle for ruled surfaces over $\mathbb{P}^1$: we shall discuss about it on Subsection 1.4.4.

1.2.4. Intersection numbers

In this subsection we shall recall a few properties of intersection numbers. See [56, Chapter IV] for proofs in dimension $n = 2$ and further details; the extension to higher dimensions is straightforward.

Definition 1.2.22. Let X be a complex n-manifold, and $D_1, \dots, D_n$ some divisors. We shall denote by $D_1 \cdot \dots \cdot D_n \in \mathbb{Z}$ the **intersection number** between $D_1, \dots, D_n$.

Remark 1.2.23. Intersection numbers are invariant with respect to linear equivalence (see [56, Chapter IV, Section 1.3 and 1.4]).

Proposition 1.2.24 ([56, Chapter IV, Section 3.2, Theorem 2.(i)]). *Let $\pi : Y \to X$ be a regular birational map between n-varieties. Let $D_1, \dots, D_n$ be divisors in X. Then*

$$\pi^*(D_1) \cdot \ldots \cdot \pi^*(D_n) = D_1 \cdot \ldots \cdot D_n.$$

Proposition 1.2.25 ([56, Chapter IV, Section 3.2, Theorem 2.(ii)]). *Let $\pi : Y \to X$ be a regular birational map between n-varieties, and denote by E the exceptional divisor and by L the indeterminacy set of π^{-1}. Let $D_1, \dots, D_k$ be divisors in X, and $E_1, \dots, E_{n-k}$ be divisors in Y. Assume that $1 \le k \le n-1$, that there is at least one of the E_j's whose components are contained in E, and that one of the D_i's is linearly equivalent to a divisor that does not intersect L. Then*

$$\pi^*(D_1) \cdot \ldots \cdot \pi^*(D_k) \cdot E_1 \cdot \ldots \cdot E_{n-k} = 0.$$

Remark 1.2.26. If the map π in Proposition 1.2.25 is such that the indeterminacy set L of π^{-1} consists in a finite number of points, then every D_i as in the statement of the Proposition is linearly equivalent to a divisor that does not intersect L (see [56, Chapter III, Section 3.1, Theorem 1]), so the last condition is always fulfilled. For example, it happens when π is a composition of blow-ups, whose centers are in the exceptional divisor of the previous blow-up, and the first one is over a point of X.

Proposition 1.2.27. *Let X be an n-manifold, and let $D_1, \dots, D_n$ be divisors in X. Then*

$$D_1 \cdot \ldots \cdot D_n = (D_1 \cdot \ldots \cdot D_{n-1})|_{D_n}.$$

Corollary 1.2.28. *Let X be a surface, C a curve in it, and $\pi : \widetilde{X} \to X$ the blow-up of X in a smooth point p of C. Let us denote by E the exceptional divisor, and by $\widetilde{C}$ the strict transform of C. Then*

$$E \cdot E = -1, \qquad \widetilde{C} \cdot \widetilde{C} = C \cdot C - 1.$$

Proof. Since p is a smooth point in C, then $\pi^*C = \widetilde{C} + E$ and $\widetilde{C}$ meets transversely E in a point, and hence $\widetilde{C} \cdot E = 1$. Hence applying Proposition 1.2.25 we get

$$0 = \pi^*C \cdot E = (\widetilde{C} + E) \cdot E = 1 + E \cdot E$$

and hence $E \cdot E = -1$. Applying then Proposition 1.2.24 we get

$$C \cdot C = \pi^*C \cdot \pi^*C = (\widetilde{C} + E) \cdot (\widetilde{C} + E) = \widetilde{C} \cdot \widetilde{C} + 2\widetilde{C} \cdot E + E \cdot E = \widetilde{C} \cdot \widetilde{C} + 1,$$

and we are done. □

Corollary 1.2.29. *Let X be a 3-manifold, and D, E be two divisors in X. Then we have*

$$\begin{aligned} D \cdot D \cdot E &= D|_E \cdot D|_E \\ &= D|_D \cdot E|_D = \mathcal{N}_{D\subset X}\, E|_D. \end{aligned}$$

In particular if $D = E$ we get

$$D \cdot D \cdot D = D|_D \cdot D|_D = \mathcal{N}_{D\subset X} \cdot \mathcal{N}_{D\subset X}.$$

1.2.5. Valuations

Definition 1.2.30. Let R be a ring (commutative, with unity), K its quotient field, and Γ an abelian totally ordered group. A **Krull valuation** is a map $\nu : K^* \to \Gamma$ such that:

(i) $\nu(\phi\psi) = \nu(\phi) + \nu(\psi)$ for all $\phi, \psi \in K^*$;
(ii) $\nu(\phi + \psi) = \min\{\nu(\phi), \nu(\psi)\}$ for all $\phi, \psi \in K^*$;
(iii) $\nu(1) = 0$.

We shall call ν **centered** if $\nu(\mathfrak{m}) > 0$. We will call $\nu(K^*)$ (resp. $\nu(R^*)$) the **value group** (resp. **value semi-group**) of ν.

Two Krull valuations $\nu_1 : K^* \to \Gamma_1$ and $\nu_2 : K^* \to \Gamma_2$ are **equivalent** ($\nu_1 \sim \nu_2$) if there exists a strictly increasing homomorphism $h : \Gamma_1 \to \Gamma_2$ such that $h \circ \nu_1 = \nu_2$.

Remark 1.2.31. Let R be a ring (commutative, with unity), and K its quotient field. If $\nu : R^* \to \Gamma$ is a map such that (i), (ii) and (iii) of Definition 1.2.30 holds, then there exists an unique valuation $\nu : K^* \to \Gamma$ that coincides with the given map on R^*.

Definition 1.2.32. Let K be a field (the main example will be $K = \mathbb{C}((x, y))$). A **valuation ring** over K is a local ring S with quotient field K and such that if $x \in K^*$ then $x \in S$ or $x^{-1} \in S$.

We can associate a valuation ring to every Krull valuation: if $\nu : K \to \Gamma$ is a Krull valuation, then $R_\nu := \{\nu \geq 0\} \cup \{0\}$ is a valuation ring over K (with maximal ideal $\{\nu > 0\}$). Vice versa, from a valuation ring S over K we can construct a Krull valuation ν_S as follows: we say that $x \sim_S y$ if there exists u unit in S such that $x = uy$. Then the projection $\nu_S : K^* \to K^*/\sim_S$ defines a Krull valuation. If we consider Krull valuations up to equivalence, this correspondence is 1-to-1, and centered Krull valuations correspond to valuation rings S such that $R \cap \mathfrak{m}_S = \mathfrak{m}$ (with $\mathfrak{m}_S$ the maximal ideal of S).

Definition 1.2.33. Let $\nu : K^* \to \Gamma$ be a Krull valuation. We define:

- the **rank** $\operatorname{rk} \nu$ of ν as the Krull dimension of R_ν;
- the **rational rank** $\operatorname{ratrk} \nu$ of ν as $\operatorname{ratrk} \nu := \dim_{\mathbb{Q}}(\nu(K^*) \otimes_{\mathbb{Z}} \mathbb{Q})$;
- the **transcendence degree** $\operatorname{trdeg} \nu$ of ν as the transcendence degree of $k_\nu := R_\nu/\mathfrak{m}_\nu$ over $\mathbb{C}$.

For proofs and further details on valuations, see [64, Part VI].

1.3. Algebraic topology

In this section we shall just recall a few classical results on homotopy, homology and cohomology groups. For references, see [36].

Definition 1.3.1. Let X be a topological space. We shall denote by $\pi_1(X)$ the **first homotopy group** of X.

Theorem 1.3.2 (Van Kampen Theorem). *Let X be a (path-connected) topological space, decomposed as $X = A \cup B$ where A and B are two path-connected open subsets of X such that $A \cap B$ is path-connected. Let us consider the first fundamental groups with respect to a base point $x_0 \in A \cap B$. Then*

$$\pi_1(X) \cong \big(\pi_1(A) * \pi_1(B)\big)/N,$$

where $*$ denotes the free product, and N is the normal subgroup generated by $i_A(g)i_B(g)^{-1}$ for every $g \in \pi_1(A \cap B)$, where $i_A : \pi_1(A \cap B) \to \pi_1(A)$ and $i_B : \pi_1(A \cap B) \to \pi_1(B)$ are the homomorphisms induced by the inclusions.

Definition 1.3.3. Let X be a topological space. We shall denote by $H_k(X)$ the k**-homology group** (with coefficients in $\mathbb{Z}$), and by $H^k(X)$ the k-**cohomology group**. We shall denote by $\widetilde{H}_0(X)$ (resp., $\widetilde{H}^0(X)$) the **reduced** 0-homology (resp. 0-cohomology) group.

Definition 1.3.4. Let X be a topological space. For every $k \in \mathbb{N}$, we shall denote by b_k the dimension of $H_k(X)$ the k-homology group. The number b_k is called k**-th Betti number**.

We shall see now a very powerful tool to compute Homology and Cohomology groups, the Mayer Vietoris Sequence (in Homology and Cohomology, see [36, pp: 149–153]).

Theorem 1.3.5 (Mayer Vietoris Sequence in Homology). *Let X be a topological space, and $A, B \subset X$ two subspaces such that X is the union* *of the interiors of A and B. Then we have the long exact sequence*

$$
\begin{array}{ccccccccc}
 & \vdots & & \vdots & & \vdots & & \vdots & \\
\to & H_n(A \cap B) & \to & H_n(A) & \oplus & H_n(B) & \to & H_n(A \cup B) & \to \\
\to & H_{n-1}(A \cap B) & \to & H_{n-1}(A) & \oplus & H_{n-1}(B) & \to & H_{n-1}(A \cup B) & \to \\
 & \vdots & & \vdots & & \vdots & & \vdots & \\
\to & H_0(A \cap B) & \to & H_0(A) & \oplus & H_0(B) & \to & H_0(A \cup B) & \to 0.
\end{array}
$$

The Mayer Vietoris Sequence holds also for the reduced homology, hence by replacing $\widetilde{H}_0$ instead of H_0.

Theorem 1.3.6 (Mayer Vietoris Sequence in Cohomology). *Let X be a topological space, and $A, B \subset X$ two subspaces such that X is the union of the interiors of A and B. Then we have the long exact sequence*

$$
\begin{array}{ccccccccc}
0 \to & H^0(A \cup B) & \to & H^0(A) & \oplus & H^0(B) & \to & H^0(A \cap B) & \to \\
\to & H^1(A \cup B) & \to & H^1(A) & \oplus & H^1(B) & \to & H^1(A \cap B) & \to \\
 & \vdots & & \vdots & & \vdots & & \vdots & \\
\to & H^n(A \cup B) & \to & H^n(A) & \oplus & H^n(B) & \to & H^n(A \cap B) & \to \\
 & \vdots & & \vdots & & \vdots & & \vdots & .
\end{array}
$$

The Mayer Vietoris Sequence holds also for the reduced cohomology, hence by replacing $\widetilde{H}^0$ instead of H^0.

Theorem 1.3.7 (Poincaré Duality, [36, Theorem 3.30]). *Let X be a closed orientable n-manifold. Then*

$$H_k(X) \cong H^{n-k}(X)$$

for all $k = 0, \ldots, n$.

1.4. Compact complex varieties

1.4.1. Minimal models

Definition 1.4.1. Here we shall consider n-manifolds, *i.e.*, compact complex smooth manifolds of complex dimension n. In this setting, we shall refer to 1-manifolds as **curves**, and to 2-manifolds as **surfaces**.

The first step that has been found to be successful to classify 2-manifolds was to reduce ourselves to surfaces that do not arise from others by blowing-up: these surfaces are called "minimal models".

Definition 1.4.2. A n-manifold X is called **minimal model** if does not exist a n-manifold Y and a modification $\pi : X \to Y$.

Theorem 1.4.3 (Castelnuovo-Enriques Criterion, [31, p.476]). *Let X be a 2-manifold, and $D \subset X$ a rational curve of self-intersection -1. Then there exists a 2-manifold Y such that X is the total space of the blow-up $\pi : X \to Y$ of a point $p \in Y$, and $D = \pi^{-1}(p)$.*

Remark 1.4.4. Clearly also the converse of Theorem 1.4.3, since the exceptional divisor of a point blow-up in dimension 2 has always self-intersection -1. Hence in dimension 2 we can define minimal model as a 2-manifold that does not have rational curves with self-intersection -1.

1.4.2. Kodaira Dimension

Then the second step for a modern exposition of the classification of surfaces is to divide them with respect to a birational invariant, called "Kodaira dimension".

Definition 1.4.5. Let X be a n-manifold. For every $m \in \mathbb{N}^*$, we shall call m-th **plurigenera** the dimension

$$P_m := h^0\big(X, \mathcal{O}\,(m\mathcal{K}_X)\big)$$

of the space of holomorphic sections of the line bundle $m\mathcal{K}_X$, where $\mathcal{K}_X$ is the canonical bundle of X.

Definition 1.4.6. Let X be a n-manifold. The **Kodaira dimension** of X is

$$\text{kod}\,(X) = \min\{k \mid P_m = O(m^k) \text{ for } m \to +\infty\}.$$

The Kodaira dimension tells us somehow how much the canonical bundle is ample.

When $P_m = 0$ for every $m \geq 1$, with this definition, the Kodaira dimension is naturally defined to be $-\infty$. There are other definitions for the Kodaira dimension, where this case is called of Kodaira dimension -1. In the literature, both notations are used.

Now we shall state some properties of the Kodaira dimension (see [11, Chapter I, Section 7]).

Proposition 1.4.7. *Let X be a n-manifold. Then the Kodaira dimension* kod (X) *of X is a birational invariant, and*

$$\text{kod}\,(X) \in \{-\infty, 0, \ldots, n\}.$$

Definition 1.4.8. Let X be a n-manifold. The **canonical graded ring** of X is the graded ring

$$R_X := \bigoplus_{m=0}^{\infty} H^0\big(X, \mathcal{O}\,(m\mathcal{K}_X)\big).$$

Proposition 1.4.9. *Let X be a n-manifold. Then*

$$\mathrm{kod}\,(X) = \mathrm{trdeg}_{\mathbb{C}}\, R_X - 1,$$

where $\mathrm{trdeg}_{\mathbb{C}}$ *denotes the transcendence degree over* $\mathbb{C}$.

In dimension 1, the Kodaira dimension can be easily computed, obtaining the following result (see [5, Example VII.2]).

Proposition 1.4.10. *Let X be a* 1*-manifold,* i.e., *a compact complex Riemann surface. If $g(X)$ is its genus, then*

- $\mathrm{kod}\,(X) = -\infty$ *if and only if* $g(X) = 0$, *and hence* $X \cong \mathbb{P}^1$.
- $\mathrm{kod}\,(X) = 0$ *if and only if* $g(X) = 1$, *and hence X is a complex torus.*
- $\mathrm{kod}\,(X) = 1$ *if and only if* $g(X) \geq 2$.

Definition 1.4.11. A n-manifold X with Kodaira dimension $\mathrm{kod}\,(X) = n$ is called of **general type**.

In dimension 2, surfaces of general type are not completely understood. There are results on the structure of the moduli spaces, but it seems not easy to compute them for all the cases.

However for Kodaira dimension 1 and 0 the classification is done and classical, while for Kodaira dimension $-\infty$, only one case is still not completely understood.

1.4.3. Class VII

Definition 1.4.12. A surface X is called of **class** VII if it has Kodaira dimension $\mathrm{kod}\,(X) = -\infty$ and first Betti number $b_1 = 1$. If moreover X is a minimal model, it is called of **class** VII_0.

In particular, class VII surfaces are not Kähler, thanks to the next proposition (see for example [62, Corollary 5.2]).

Proposition 1.4.13. *Let X be a compact Kähler manifold. Then odd Betti numbers are even.*

Then, when the second Betti number $b_2 = 0$, these surfaces have been completely classified, thanks to the work of Kodaira ([42, 43]), Inoue ([38]), Bogomolov ([8]), Li, Yau and Zheng ([48]), Teleman ([60]).

For $b_2 > 0$, the classification is not completed yet. Before describing the known results, we need a definition.

Definition 1.4.14. Let X be a n-manifold. A **spherical shell** is a holomorphic embedding $i : V \hookrightarrow X$, where V is a neighborhood of $\mathbb{S}^{2n-1} = \partial\mathbb{B}^{2n} \subset \mathbb{C}^n$. A spherical shell is said to be **global** (or **GSS**) if $X \setminus i(V)$ is connected; it is said to be **local** otherwise.

Remark 1.4.15. While any n-manifold admits a local spherical shell, the existence of a global spherical shell is not always possible. For example if $n = 1$, a global spherical shell exists only when $g \geq 1$, where g denotes the genus.

Kato introduced a construction method for surfaces of class VII_0 with $b_2 > 0$, called **Kato surfaces** (see [40]). For these surfaces, the following properties hold.

Theorem 1.4.16 (Kato). *Let X be a surface of class VII_0 with $b_2 = b_2(X) > 0$. Then X admits at most b_2 rational curves. If moreover X admits a GSS, then X has exactly b_2 rational curves.*

Kato conjectured that the converse should also be true. This result has been proved by Dloussky, Oeljeklaus and Toma (see [20]).

Theorem 1.4.17. *If X is a surface of class VII_0 with $b_2 = b_2(X) > 0$ and with b_2 rational curves, then X admits global spherical shells.*

There are not known examples of surfaces X of class VII_0 that do not admit global spherical shells, and it has been conjectured that there are none.

1.4.4. Ruled Surfaces

In the following we shall need some knowledge about another class of surfaces which arise as exceptional divisors of blow-up (of curves).

Definition 1.4.18. A surface X is called **ruled surface** over a curve C if it is the total space of a fiber bundle $\pi : X \to C$ with projective lines $\mathbb{P}^1$ as fibers. If moreover $C \cong \mathbb{P}^1$ is a rational curve, then X is called **rational**.

The next proposition tells us that all fiber bundles can be taken as projectivization of vector bundles. For proofs, see [5, Proposition III.7].

Proposition 1.4.19. *Every ruled surface over a curve C is isomorphic to the projectivization $\mathbb{P}(E)$ of a holomorphic vector bundle E of rank 2 over C. Two such surfaces $\mathbb{P}(E)$ and $\mathbb{P}(F)$ are isomorphic if and only if there exists a line bundle L over C such that $E \cong F \otimes L$.*

As we shall see, ruled surfaces are surfaces for which the minimal model is not unique. For the proof of the next proposition, see [5, Chapter III].

Proposition 1.4.20. *All ruled surfaces over a curve C are birationally equivalent to $C \times \mathbb{P}^1$.*

Definition 1.4.21. In the following, we shall need only to blow-up rational curves, and hence we shall see only rational ruled surfaces.

The following proposition tells us the structure of holomorphic line bundles over $\mathbb{P}^1$. For proofs, see [5, Proposition III.15], or [31, p. 516].

Proposition 1.4.22. *Any holomorphic line bundle E of rank r over $\mathbb{P}^1$ is decomposable,* i.e., *$E \cong L_1 \oplus \ldots \oplus L_r$ for suitable line bundles L_i.*

Definition 1.4.23. For every $k, h \in \mathbb{Z}$, we shall denote by $\mathbb{F}_{k,h}$ the rational ruled surface

$$\mathbb{F}_{k,h} := \mathbb{P}(\mathcal{O}(k) \oplus \mathcal{O}(h)).$$

We shall usually denote by H the fiber of a point, $L_0 = \mathbb{P}(\mathcal{O}(k) \oplus 0)$ and $L_\infty = \mathbb{P}(0 \oplus \mathcal{O}(h))$, considered as divisors of $\mathbb{F}_{k,h}$.

For every $n \geq 0$, we shall denote by $\mathbb{F}_n$ the rational ruled surface

$$\mathbb{F}_n := \mathbb{P}(\mathcal{O}(n) \oplus \mathcal{O}(0)) = \mathbb{F}_{n,0}. \tag{1.1}$$

Remark 1.4.24. In the literature, the notation $\mathbb{F}_n := \mathbb{F}_{0,n}$ is perhaps more used. Given a chart $\mathbb{C} \subset \mathbb{P}^1$, the ruling over $\mathbb{C}$ is trivial, and we have $\mathbb{C} \times \mathbb{P}^1$. For every $c \in \mathbb{P}^1$ we can consider a section $L_c = \{(z, c) \in \mathbb{C} \times \mathbb{P}^1\}$ and extend it in $\mathbb{F}_n$. Then the notation is consistent, *i.e.*, L_0 and L_∞ defined here are the same as in Definition 1.4.23. We chose notations in order to have L_∞ to be the "special curve" with respect to other curves L_c, $c \in \mathbb{C}$, see Proposition 1.4.26.

Corollary 1.4.25. *Any rational ruled surface over $\mathbb{P}^1$ is isomorphic to $\mathbb{F}_n$ for a suitable $n \geq 0$. The rational ruled surface $\mathbb{F}_{k,h}$ is isomorphic to $\mathbb{F}_n$ if and only if $n = |k - h|$.*

Proposition 1.4.26 ([5, Proposition IV.1]). *Let $\mathbb{F}_n$ be the rational ruled surface defined by* (1.1). *Then*

(i) $\mathrm{Pic}(\mathbb{F}_n) \cong \mathbb{Z}H \oplus \mathbb{Z}L_0$, *where $H \cdot H = 0$, $H \cdot L_0 = 1$ and $L_0 \cdot L_0 = n$.*
(ii) *If $n > 0$, then L_∞ is the unique curve with negative self intersection, $L_\infty = L_0 - nH$, and hence $H \cdot L_\infty = 1$ and $L_\infty \cdot L_\infty = -n$.*
(iii) *$\mathbb{F}_n \cong \mathbb{F}_m$ if and only if $n = m$. $\mathbb{F}_n$ is a minimal model for every $n \neq 1$. $\mathbb{F}_1$ is isomorphic to $\mathbb{P}^2$ blown-up once in a point.*

Remark 1.4.27. We can explicitly construct a birational map between rational ruled surfaces.

Let $\mathbb{F}_n$ be a rational ruled surface, and H, L_0, L_∞ as in Definition 1.4.23. We take X the blow-up of $\mathbb{F}_n$ at a point p in a fiber H. Thanks to Corollary 1.2.28, we have that $\widetilde{H} \cdot \widetilde{H} = -1$, where $\widetilde{H} \cong \mathbb{P}^1$ is the strict transform of H; moreover, if we denote by E the exceptional divisor, we have $E \cdot E = -1$. Finally, if $p \in L_\infty$, then the strict transform $\widetilde{L_\infty}$ of L_∞ is such that $\widetilde{L_\infty} \cdot \widetilde{L_\infty} = -n - 1$.

Thanks to Theorem 1.4.3, we can then blow-down $\widetilde{H}$, obtaining a surface Y. We shall denote by E' and L'_∞ the projections of E and $\widetilde{L_\infty}$ respectively. If $p \in L_\infty$, then $\widetilde{L_\infty}$ does not intersect $\widetilde{H}$, and hence $L'_\infty \cdot L'_\infty = -n-1$; otherwise, we have $L'_\infty \cdot L'_\infty = -n+1$. In both cases $E' \cdots E' = 0$. If we denote by $\mathrm{pr} : \mathbb{F}_n \to L_\infty$ the projection given by the ruled surface structure, then we can define $\mathrm{pr}' : Y \to L'_\infty$ by $\mathrm{pr}' = \mathrm{pr}$ outside E', and $\mathrm{pr}'(E)$ equal to the point in $L'_\infty \cap E'$.

Then Y is a ruled surface, and thanks to Corollary 1.4.25 and Proposition 1.4.26, we have that $Y \cong \mathbb{F}_{n+1}$ if $p \in L_\infty$, and $Y \cong \mathbb{F}_{n-1}$ if $p \notin L_\infty$ (if $n > 0$, while $Y \cong \mathbb{F}_1$ if $n = 0$).

Finally, we shall need the computation on the canonical bundle for rational ruled surfaces (see for example [5, Proposition III.18]).

Proposition 1.4.28. *Let $\mathbb{F}_n$ be the ruled surface, and H, L_0, L_∞ the divisors in it defined as in Definition* 1.4.23.

Then the canonical bundle is

$$\mathcal{K}_{\mathbb{F}_n} = (-2-n)[H] - 2[L_\infty] = (-2+n)[H] - 2[L_0].$$

Chapter 2
Dynamics in 2D

2.1. The valuative tree

2.1.1. Tree structure

In this section we shall work with a given totally ordered set Λ (typical examples of Λ are $\mathbb{R}$ or $\mathbb{Q}$ or $\mathbb{N}$), that will be the model of the tree.

Definition 2.1.1. Let $(\mathcal{T}, \leq)$ be a poset; then a totally ordered subset $\mathcal{S} \subseteq \mathcal{T}$ is **full** if for every $\sigma_1, \sigma_2 \in \mathcal{S}$ and $\tau \in \mathcal{T}$ such that $\sigma_1 \leq \tau \leq \sigma_2$, then $\tau \in \mathcal{S}$.

Definition 2.1.2. Let $(\mathcal{T}, \leq)$ be a poset; it is a (non-metric) Λ**-tree** if:

(T1) there exists an unique minimal element $\tau_0 \in \mathcal{T}$, called the **root** of $\mathcal{T}$;

(T2) for every $\tau \in \mathcal{T}$, the set $\{\sigma \in \mathcal{T} \mid \sigma \leq \tau\}$ is isomorphic (as ordered sets) to an interval of Λ;

(T3) every totally ordered full subset $\mathcal{S} \subseteq \mathcal{T}$ is isomorphic (as ordered sets) to an interval of Λ.

The third condition is equivalent to:

(T4) for every $\mathcal{S} \subseteq \mathcal{T}$, without upper bounds in $\mathcal{T}$, there exists an increasing (countable) sequence in $\mathcal{S}$ without upper bounds in $\mathcal{T}$.

Maximal elements of $\mathcal{T}$ are called **ends**.

A Λ-tree is **complete** if every increasing (countable) sequence has an upper bound in $\mathcal{T}$.

Now we fix $\Lambda = \mathbb{R}$; using the completeness of $\mathbb{R}$, we can define more objects in an $\mathbb{R}$-tree.

Definition 2.1.3. Let $(\mathcal{T}, \leq)$ be an $\mathbb{R}$-tree. For every subset $\mathcal{S} \subseteq \mathcal{T}$, we can define $\bigwedge_{\sigma \in \mathcal{S}} \sigma$ as the infimum over the elements in $\mathcal{S}$ (it exists thanks to the completeness of $\mathbb{R}$). Then we define the **(closed) segment** between τ_1 and τ_2 as:

$$[\tau_1, \tau_2] := \{\tau \in \mathcal{T} \mid \tau_1 \wedge \tau_2 \leq \tau \leq \tau_1 \text{ or } \tau_1 \wedge \tau_2 \leq \tau \leq \tau_2\},$$

and analogously the semi-open and open segments $[\tau_1, \tau_2), (\tau_1, \tau_2)$.

We can define tangent vectors on a point of an $\mathbb{R}$-tree.

Definition 2.1.4. Let $(\mathcal{T}, \leq)$ be an $\mathbb{R}$-tree, and $\tau \in \mathcal{T}$ a point. We say that $\sigma_1, \sigma_2 \in \mathcal{T} \setminus \{\tau\}$ are equivalent ($\sigma_1 \sim \sigma_2$) if and only if $(\tau, \sigma_1] \cap (\tau, \sigma_2] \neq \emptyset$. We call $T_\tau\mathcal{T} := \mathcal{T} \setminus \{\tau\}/\sim$ the **tangent space** of $\mathcal{T}$ over τ, and we denote by $\overrightarrow{v} = [\sigma] \in T_\tau\mathcal{T}$ a tangent vector over τ (represented by σ).

The point τ is a **terminal point**, a **regular point** or a **branch point** if $T_\tau\mathcal{T}$ has $1, 2$ or more than 2 tangent vectors respectively.

Using tangent vectors we can define a topology on an $\mathbb{R}$-tree.

Definition 2.1.5. Let $(\mathcal{T}, \leq)$ be an $\mathbb{R}$-tree; for every point $\tau \in \mathcal{T}$ and tangent vector $\overrightarrow{v} = T_\tau\mathcal{T}$, define:

$$U_\tau(\overrightarrow{v}) := \{\sigma \in \mathcal{T} \mid \overrightarrow{v} = [\sigma]\}.$$

This sets are the prebasis of a topology on $\mathcal{T}$, called the **weak topology** on $\mathcal{T}$.

Definition 2.1.6. Let $(\mathcal{T}, \leq)$ an $\mathbb{R}$-tree. For every totally ordered segment $I = [\tau_1, \tau_2]$ in $\mathcal{T}$, we can construct a **retraction** from $\mathcal{T}$ to I, defined by:

$$\pi_I(\sigma) = \begin{cases} \sigma \wedge \tau_2 & \text{if } \sigma(\wedge\tau_2) \geq \tau_1 \,, \\ \tau_1 & \text{otherwise.} \end{cases}$$

It can be easily seen that the weak topology is the weakest one for which all retractions are continuous.

Remark 2.1.7. From an $\mathbb{N}$-tree, we can easily obtain an $\mathbb{R}$-tree by “joining points”; we can do it even from a $\mathbb{Q}$-tree, completing the $\mathbb{Q}$-tree to an $\mathbb{R}$-tree with a construction similar to the completion of $\mathbb{Q}$ to $\mathbb{R}$ (see [26, Proposition 3.12]). So we can extend all these definitions from $\mathbb{R}$-trees to $\mathbb{N}$-trees and $\mathbb{Q}$-trees, considering the latter trees as embedded in $\mathbb{R}$-trees.

Definition 2.1.8. Let $(\mathcal{T}, \leq)$ be a Λ-tree; a **parametrization** of $\mathcal{T}$ is an increasing (or decreasing) map $\alpha : \mathcal{T} \to \Lambda$ whose restriction to any full, totally ordered subset of $\mathcal{T}$ gives a bijection onto an interval of Λ.

When $\Lambda = \mathbb{R}$, it is sometimes useful to consider parametrizations $\alpha : \mathcal{T} \to \overline{\mathbb{R}} = [-\infty, +\infty]$. It can be easily seen that all parametrizations are lower semi-continuous with respect to the weak topology (see [26, Proposition 3.8]).

Tree Maps

We introduced the objects, Λ-trees; now we define morphisms.

Definition 2.1.9. Let $(\mathcal{T}, \leq)$ and $(\mathcal{S}, \leq)$ be Λ-trees, with roots τ_0 and σ_0 respectively. A map $\Phi : \mathcal{T} \to \mathcal{S}$ is called a **tree morphism** if Φ induces an order preserving bijection of $[\tau_0, \tau]$ onto $[\sigma_0, \Phi(\tau)]$ for every $\tau \in \mathcal{T}$.
If Φ is also a bijection, then we call it a **tree isomorphism**.

For our purposes, we need a larger family of maps between trees than morphisms: for complete $\mathbb{R}$-trees, we can define tree maps.

Definition 2.1.10. Let $(\mathcal{T}, \leq)$ and $(\mathcal{S}, \leq)$ be two complete $\mathbb{R}$-trees, and $F : \mathcal{T} \to \mathcal{S}$ a map. F is called **tree map** if the restriction $f|_I : I \to \mathcal{S}$ is weakly-continuous for every segment $I \subseteq \mathcal{T}$. A tree map is **regular** if any segment $I \subseteq \mathcal{T}$ can be decomposed into finitely many segments on each of which F is a homeomorphism onto its image.

Remark 2.1.11. A tree map $F : \mathcal{T} \to \mathcal{T}$ naturally induces a tree map $F_I : I \to I$ for every totally ordered segment I, by setting $F_I := \pi_I \circ F|_I$, where π_I is the retraction from $\mathcal{T}$ to I.

In the next sections, we will need a fixed point theorem for tree maps; we will state the result here.

Definition 2.1.12. Let $(\mathcal{T}, \leq)$ be a complete $\mathbb{R}$-tree, and $F : \mathcal{T} \to \mathcal{T}$ a tree map. An end $\tau \in \mathcal{T}$ is **weakly attracting** for F if there exists a segment $I = [\tau', \tau]$ such that τ is a globally attracting fixed point for the induced map F_I, *i.e.*, $F_I(\sigma) > \sigma$ for all $\sigma \in [\tau', \tau)$. The end is **strongly attracting** (for F) when in addition the segment I can be chosen F-invariant.

Theorem 2.1.13 ([27, Theorem 4.5], [53]). *Let $(\mathcal{T}, \leq)$ be a complete $\mathbb{R}$-tree and $F : \mathcal{T} \to \mathcal{T}$ a tree map (resp. a regular tree map). Then at least one of the following statements holds:*

- *F admits a fixed point $\tau_\star$ which is not an end;*
- *F admits a weakly (resp. strongly) attracting end $\tau_\star$.*

2.1.2. Universal Dual Graph

Given a modification $\pi \in \mathfrak{B}$, we can equip the set Γ_π^* of all exceptional components of π with a simplicial tree structure (*i.e.*, an $\mathbb{N}$-tree structure, see [26, pages 51, 52]), given by the dual graph of π as defined in Definition 1.2.14. We shall denote the order given by the $\mathbb{N}$-tree structure by $\leq_\pi$.

From the recursive construction of the dual graph of a modification, as seen in Remark 1.2.15, we can observe that if $\pi_1 \trianglelefteq \pi_2$ then there is a natural injection of $\Gamma^*_{\pi_1}$ into $\Gamma^*_{\pi_2}$. Since $\mathfrak{B}$ is an inverse system, we can give the next definition.

Definition 2.1.14. We will call **universal dual graph** the direct limit of dual graphs along all modifications in $\mathfrak{B}$:

$$(\Gamma^*, \leq) := \varinjlim_{\pi \in \mathfrak{B}} (\Gamma^*_\pi, \leq_\pi).$$

The universal dual graph is a way to see all exceptional components of all the possible modifications, all together at the same time. The next result follows from this construction.

Proposition 2.1.15 ([26, Proposition 6.2, Proposition 6.3]). *The universal dual graph Γ^* is a $\mathbb{Q}$-tree, rooted at E_0 the exceptional component that arises from the single blow-up of the origin $0 \in \mathbb{C}^2$. Moreover all points are branch points for Γ^*. If we have an exceptional component $E \in \Gamma^*$, then $p \mapsto \overrightarrow{v_p} = [E_p]$, where $[E_p] \in T_E\Gamma^*$ is the tangent vector represented by the exceptional component that arises from the blow-up of p, gives a bijection from E to $T_E\Gamma^*$.*

As we have seen in Remark 2.1.7, we can complete Γ^* to a complete $\mathbb{R}$-tree Γ, that will also be called the (complete) **universal dual graph**.

The (complete) universal dual graph is a very powerful tool, also thanks to all the structure that arises from the completeness of $\mathbb{R}$. But we do not know how a holomorphic germ f acts on the universal dual graph. The answer to this question can be given thanks to the algebraic equivalent to the universal dual graph, the **valuative tree**.

2.1.3. Valuations

We shall denote by $R = \mathbb{C}[[x, y]]$ the ring or formal power series in 2 coordinates, and by $K = \mathbb{C}((x, y))$ the quotient field of R, that is the field of Laurent series in 2 coordinates. Then R is an UFD local ring, with maximal ideal $\mathfrak{m} = \langle x, y \rangle$. We shall consider $[-\infty, \infty] = \overline{\mathbb{R}}$ endowed with the standard extension of the usual order $\leq$ in $\mathbb{R}$.

Definition 2.1.16. A **valuation** is a map $\nu : R \to [0, \infty]$ such that:

(i) $\nu(\phi\psi) = \nu(\phi) + \nu(\psi)$ for all $\phi, \psi \in R$;
(ii) $\nu(\phi + \psi) = \min\{\nu(\phi), \nu(\psi)\}$ for all $\phi, \psi \in R$;
(iii) $\nu(1) = 0$.

Denote by $\mathfrak{p} = \{\nu = \infty\}$ the prime ideal where ν is ∞. Then ν is **proper** if $\mathfrak{p} \subsetneq \mathfrak{m}$. For every ideal $\mathfrak{i} \subset R$, we denote $\nu(\mathfrak{i}) := \min\{\nu(\phi) \mid \phi \in \mathfrak{i}\}$; then ν is **centered** if it is proper and $\nu(\mathfrak{m}) > 0$; and it is **normalized** if it is centered and $\nu(\mathfrak{m}) = 1$. We will denote by $\tilde{\mathcal{V}}$ the set of all centered valuations, and by $\mathcal{V}$ the set of all normalized ones.

Let us define a partial order on centered valuations.

Definition 2.1.17. Let ν_1, ν_2 be two centered valuations; then $\nu_1 \leq \nu_2$ if and only if $\nu_1(\phi) \leq \nu_2(\phi)$ for every $\phi \in R$.

The set of all normalized valuations will be the object of main interest: we will see that it admits an $\mathbb{R}$-tree structure, so we call $\mathcal{V}$ the **valuative tree**.

Remark 2.1.18. The definition of valuation we use is slightly different from the one of Krull valuations (where ν takes values into an abelian totally ordered group Γ, see Definition 1.2.30). Indeed we can associate a Krull valuation to every proper valuation. As a matter of fact, given a proper valuation ν, if $\mathfrak{p} = \{\nu = \infty\}$, then we can have two cases:

$\mathfrak{p} = 0$, and ν is a Krull valuation (with $\Gamma = \mathbb{R}$);
$\mathfrak{p} = \langle\phi\rangle$ for a certain ϕ irreducible; in this case, we can define $\mu = \mathrm{Krull}(\nu) : R^* \to \mathbb{R} \times \mathbb{Z}$, as following: if $\psi = \phi^n \psi'$, with ψ' and ϕ coprime, then $\mu(\psi) := (\nu(\psi'), n)$.

On the other hand, there exist centered Krull valuations in $\mathbb{C}[[x, y]]$ that do not arise from proper valuations (they are called exceptional curve valuations, see the next subsection below for further details).

However, thanks to this inclusion, we can use all invariants and tools known for the study of Krull valuations, as for instance the correspondence with valuation rings, or invariants such as the rank, the rational rank, or the transcendence degree of a Krull valuation.

2.1.4. Classification of Valuations

We shall now describe the classification of valuations, following [26]. Most of the properties stated here will be clearer after reading the rest of this section, when we will describe the tree structure of the valuative tree. For more details see also [58] and [64].

Divisorial Valuations

Divisorial valuations are the ones with $\mathrm{rk}\,\nu = \mathrm{ratrk}\,\nu = \mathrm{trdeg}\,\nu = 1$. They are associated to an exceptional component E of a modification π; in particular ν_E is defined by

$$\nu_E(\phi) := (1/b_E)\,\mathrm{div}_E(\pi^*\phi),$$

where div_E is the vanishing order along E, $\pi^*\phi = \phi \circ \pi$ and $1/b_E$ is necessary to have a normalized valuation ($b_E \in \mathbb{N}^*$ is known as the **generic multiplicity** of ν_E, see [26, page 64], or the second Farey weight of E, see [26, page 122]). The set of all divisorial valuations is often denoted by $\mathcal{V}_{\mathrm{div}}$.

The most important example is the **multiplicity valuation**, defined by

$$\nu_{\mathfrak{m}}(\phi) := m(\phi) = \max\{n \mid \phi \in \mathfrak{m}^n\};$$

it is associated to a single blow-up over the origin. We will write $\nu_{\mathfrak{m}}$ if we want to consider the multiplicity as a valuation (or better, as a point on the valuative tree), and m if we want to consider only the multiplicity of an element of $R = \mathbb{C}[[x, y]]$.

Irrational Valuations

Irrational valuations are the ones with $\operatorname{rk} \nu = 1$, $\operatorname{ratrk} \nu = 2$ and $\operatorname{trdeg} \nu = 0$.

Divisorial and irrational valuations are called **quasimonomial valuations**, their set will be denoted by $\mathcal{V}_{qm}$. For a geometric interpretation of quasimonomial valuations, see [26, pages 16, 17].

Important examples of quasimonomial valuations are **monomial valuations**. Fix local coordinates (x, y); then the monomial valuation of weights (s, t) is defined by

$$\nu_{s,t}\left(\sum_{i,j} a_{i,j} x^i y^j\right) = \min\{si + tj \mid a_{i,j} \neq 0\}.$$

They are normalized when $\min\{s, t\} = 1$.

Curve Valuations

Curve valuations are the ones with $\operatorname{rk} \nu = \operatorname{ratrk} \nu = 2$ and $\operatorname{trdeg} \nu = 0$.

They are associated to a (formal) irreducible curve (germ) $C = \{\psi = 0\}$; in particular ν_C is defined by

$$\nu_C(\phi) := \frac{C \cdot \{\phi = 0\}}{m(C)},$$

where with $C \cdot D$ we denote the standard intersection multiplicity between the curves C and D, and $m(C) = m(\psi)$ is the multiplicity of C (in 0). We will often use the notation ν_ψ instead of ν_C.

Analytic and non-analytic curve valuations have the same algebraic behavior, but they will play a different role as eigenvaluations, as we shall see in the proof of Theorem 2.3.2.

Infinitely Singular Valuations

Infinitely singular valuations are the ones with $\mathrm{rk}\,\nu = \mathrm{ratrk}\,\nu = 1$ and $\mathrm{trdeg}\,\nu = 0$.

It is not so simple to give a geometric interpretation of infinitely singular valuations, but we can think them as curve valuations associated to "curves" of infinite multiplicity.

They can be recognized also as valuations with infinitely generated value groups.

Exceptional Curve Valuations

As anticipated before, there are also Krull valuations that does not come from proper valuations; they are called **exceptional curve valuations**, and they have the same invariants as curve valuations: $\mathrm{rk}\,\nu = \mathrm{ratrk}\,\nu = 2$ and $\mathrm{trdeg}\,\nu = 0$. See [26, page 18] for more details.

2.1.5. The Valuative Tree

Theorem 2.1.19 ([26, Theorem 3.14]). *$(\mathcal{V}, \leq)$ is a complete $\mathbb{R}$-tree, with root $\nu_{\mathfrak{m}}$. We will call $(\mathcal{V}, \leq)$ the* **valuative tree***.*

Proposition 2.1.20 ([26, Proposition 3.20]). *The valuative tree $(\mathcal{V}, \leq)$ has the following properties:*

(i) *branch points are the divisorial valuations, and the tangent space on every divisorial valuation is in bijection with $\mathbb{P}^1(\mathbb{C})$;*
(ii) *regular points are the irrational valuations;*
(iii) *all terminal points are ends (and vice versa, the converse is always true), and they are curve or infinitely singular valuations.*

2.1.6. Skewness, multiplicity and thinness

The valuative tree admits (at least) two natural parametrizations (skewness and thinness) and a concept of multiplicity, very useful for example to distinguish the type of valuations. For definitions and properties we refer to [26, Chapter 3].

Definition 2.1.21. For $\nu \in \mathcal{V}$, we define its **skewness** $\alpha(\nu) \in [1, \infty]$ as

$$\alpha(\nu) := \sup\left\{\frac{\nu(\phi)}{m(\phi)} \ : \ \phi \in \mathfrak{m}\right\}.$$

Proposition 2.1.22 ([26, Theorem 3.26]). *The skewness $\alpha : \mathcal{V} \to [1,\infty]$ is a parametrization for the valuative tree. Moreover:*

(i) *the multiplicity valuation is the only one with $\alpha(\nu_{\mathfrak{m}}) = 1$;*
(ii) *for divisorial valuations we have $\alpha(\nu_E) \in \mathbb{Q}$;*

(iii) *for irrational valuations we have* $\alpha(\nu) \in \mathbb{R} \setminus \mathbb{Q}$;
(iv) *for curve valuations we have* $\alpha(\nu_C) = \infty$;
(v) *for infinitely singular valuations we have* $\alpha(\nu) \in (1, \infty]$.

Skewness alone does not characterize the type of a valuation: we need a way for telling infinitely singular valuations apart from the others. For this purpose, we introduce the multiplicity of a valuation.

Definition 2.1.23. Let $\nu \in \mathcal{V}$ be a normalized valuation. We can define its **multiplicity** $m(\nu)$ as

$$m(\nu) = \min\left\{m(\phi) \mid \nu_\phi \geq \nu,\ \phi \text{ irreducible curve}\right\}.$$

Proposition 2.1.24 ([26, Proposition 3.37]). *If* $\mu \leq \nu$ *then* $m(\mu)$ *divides* $m(\nu)$. *Further* $m(\nu) = \infty$ *if and only if* ν *is infinitely singular.*

The next proposition tells us something about the behavior of multiplicity in a segment $[\nu_{\mathfrak{m}}, \nu]$, with $\nu \in \mathcal{V}$.

Proposition 2.1.25 ([26, Proposition 3.44]). *For any valuation* $\nu \in \mathcal{V}$, *there exists a unique sequence of divisorial valuations* $(\nu_i)_{i=0}^g$ *and a unique strictly increasing sequence of integers* $(m_i)_{i=0}^g$ *such that*

$$\nu_{\mathfrak{m}} = \nu_0 < \nu_1 < \ldots < \nu_g < \nu_{g+1} = \nu,$$

and $m(\mu) = m_i$ *for* $\mu \in (\nu_i, \nu_{i+1}]$, *for* $i = 0, \ldots, g$. *Moreover* $m_0 = 1$, *while* $m_i \mid m_{i+1}$ *for* $i = 1, \ldots g - 1$.

Definition 2.1.26. We call the sequence $(\nu_i)_{i=0}^g$ the **approximating sequence** of a valuation ν. We remark that $g = \infty$ if and only if $m(\nu) = \infty$, if and only if ν is an infinitely singular valuation.

Putting together skewness and multiplicity, we obtain thinness.

Definition 2.1.27. Let $\nu \in \mathcal{V}$ be a valuation with approximating sequence $(\nu_i)_{i=0}^g$; we define the **thinness** $A(\nu) \in [2, \infty]$ of ν as

$$A(\nu) := 2 + \sum_{i=0}^{g} m_i(\alpha(\nu_{i+1}) - \alpha(\nu_i)) = 2 + \int_{\nu_{\mathfrak{m}}}^{\nu} m(\mu) d\alpha(\mu),$$

where α is the skewness, m the multiplicity, and m_i are defined as in Proposition 2.1.25.

Proposition 2.1.28 ([26, Theorem 3.46]). *The thinness* $A : \mathcal{V} \to [2, \infty]$ *is a parametrization for the valuative tree. Moreover:*

(i) *the multiplicity valuation is the only one with* $A(\nu_{\mathfrak{m}}) = 2$;
(ii) *for divisorial valuations we have* $A(\nu_E) \in \mathbb{Q}$;
(iii) *for irrational valuations we have* $A(\nu) \in \mathbb{R} \setminus \mathbb{Q}$;
(iv) *for curve valuations we have* $A(\nu_C) = \infty$;
(v) *for infinitely singular valuations we have* $A(\nu) \in (2, \infty]$.

2.1.7. Universal dual graph and valuative tree

Here we shall clarify the connection between the universal dual graph and the valuative tree. For stating it, we need the following definition.

Center of a Valuation

Definition 2.1.29. Let X be a complex 2-manifold, and V an irreducible subvariety in X. We shall denote by $\mathcal{O}_{X,V}$ the ring of regular functions in V.

Definition 2.1.30. Let $(R_1, \mathfrak{m}_1)$ and $(R_2, \mathfrak{m}_2)$ be two local rings with the same quotient field. Then we shall say that $(R_1, \mathfrak{m}_1)$ **dominates** $(R_2, \mathfrak{m}_2)$ if $R_1 \supseteq R_2$ and $R_2 \cap \mathfrak{m}_1 = \mathfrak{m}_2$.

The next theorem is a classic result of algebraic geometry: see [64, Part VI, Chapter 5], or [35] for a modern exposition.

Theorem 2.1.31 ([35, Theorem 4.7]). *Let ν be a Krull valuation on $K = \mathbb{C}((x, y)) \supset \mathbb{C}(x, y)$, R_ν the associated valuation ring and $\pi : X \to (\mathbb{C}^2, 0)$ a modification. Then there exists a unique irreducible submanifold V of X such that R_ν dominates $\mathcal{O}_{X,V}$. Moreover if ν is centered, then V is a point or an exceptional component in $\pi^{-1}(0)$.*

This V is called the **center** *of ν in X.*

The isomorphism

The center of a valuation is the main concept allowing to pass from valuations to exceptional components. The idea is to associate to a valuation, a (finite or infinite) sequence of infinitely near points, where every point is the center of the valuation in a suitable total space for a modification. This idea gives us the relation between exceptional components and (divisorial) valuations, and between the universal dual graph and the valuative tree.

Theorem 2.1.32 ([26, Theorem 6.22]). *The map $E \mapsto \nu_E$ is a tree isomorphism from the universal dual graph Γ^* to the set of all divisorial valuations $\mathcal{V}_{\mathrm{div}}$. It can be extended to a tree isomorphism from the (complete) universal dual graph Γ to the valuative tree $\mathcal{V}$.*

Favre and Jonsson prove a stronger result: they give a parametrization of the universal dual graph (the Farey parametrization, see [26, page 122]), and a concept of multiplicity of an exceptional component, and show that these objects correspond to thinness and multiplicity on the valuative tree.

Infinitely near points, best approximations and open sets

Thanks to the correspondence between the universal dual graph and the valuative tree (and to Proposition 2.1.15), we can see a correlation between infinitely near points and certain open sets of the valuative tree. But first we need to know the behavior of the center of the pull back of a valuation through a modification.

Proposition 2.1.33 ([26, Proposition 6.32]). *Let $\pi : X \to (\mathbb{C}^2, 0) \in \mathfrak{B}$ be a modification, and $\nu \in \mathcal{V}$ a valuation. Let $\mathcal{D}_\pi := \{\nu_E \mid E \in \Gamma^*_\pi\}$ be the set of all (divisorial) valuations associated to an exceptional component of π, and*

$$\mathcal{E}_{\pi,\nu} = \{\nu_E \in \mathcal{D}_\pi \mid (\nu_E, \nu) \cap \mathcal{D}_\pi = \emptyset\}.$$

Then $\mathcal{E}_{\pi,\nu}$ contains one or two valuations.

(i) *When $\mathcal{E}_{\pi,\nu} = \{\nu_E\}$, then either $\nu_E = \nu$ and the center of ν in X is E; or $\nu_E \neq \nu$, and the center of ν in X is the free point $p \in E$ associated (through Proposition 2.1.15) to the tangent vector in ν_E represented by ν.*
(ii) *If $\mathcal{E}_{\pi,\nu} = \{\nu_E, \nu_F\}$, then the center of ν in X is the satellite point $p \in E \cap F$. This point is the one in E (resp. in F) associated to the tangent vector in ν_E (resp. ν_F) represented by ν.*

Definition 2.1.34. We call the elements of $\mathcal{E}_{\pi,\nu}$ the **best approximations** of ν for π (or in Γ^*_π).

From this result we can obtain a characterization of valuations whose center is a given infinitely near point.

Proposition 2.1.35 ([26, Corollary 6.34]). *Let $p \in \pi^{-1}(0)$ be an infinitely near point of a modification $\pi : X \to (\mathbb{C}^2, 0)$; let $U(p)$ be the set of all valuations whose center in X is p. Then $U(p)$ is weakly open in $\mathcal{V}$. More precisely, let us denote by E_p the exceptional component obtained by a single blow-up over p; if p is a free point, and E is the unique exceptional component that contains p, then $U(p) = U_E([E_p])$, while if p is a satellite point, the intersection point of two exceptional components E, F, then $U(p) = U_E([E_p]) \cap U_F([E_p])$ (up to the identification between universal dual graph Γ and valuative tree $\mathcal{V}$).*

2.2. Dynamics on the valuative tree

Definition

In this section we will define the action $f_\bullet : \mathcal{V} \to \mathcal{V}$ induced by a holomorphic germ $f : (X, p) \to (Y, q)$ (where X and Y are two complex

2-manifolds). We shall also assume that f is **dominant**, *i.e.*, $\mathrm{rk}\, df$ is not identically ≤ 1 near p.

A holomorphic germ $f : (X, p) \to (Y, q)$ naturally induces an action f^* on $R = \mathbb{C}[[x, y]]$, by composition: $\phi \mapsto f^*\phi = \phi \circ f$. The natural way to define an action on (centered) valuations seems to be the dual action $f_*\nu = \nu \circ f^*$; explicitly we have $f_*\nu(\phi) = \nu(\phi \circ f)$. This definition works for Krull valuations, but not for valuations: if $\nu \in \mathcal{V}$, then clearly $f_*\nu$ is a valuation, but it might not be proper. More precisely, $f_*\nu$ is not centered if and only if $\nu = \nu_C$ is a curve valuation, with $C = \{\phi = 0\}$ an irreducible curve contracted to q by f (that is to say if $f^*\mathfrak{m} \subseteq \langle\phi\rangle$). In this case C has to be a critical curve, and $f_*\nu$ is not proper.

Definition 2.2.1. Let $f : (X, p) \to (Y, q)$ be a (dominant) holomorphic germ. We call **contracted critical curve valuations** for f the valuations ν_C with C a critical curve contracted to q by f. We denote by $\mathfrak{C}_f$ the set of all contracted critical curve valuations for f.

Remark 2.2.2. $\mathfrak{C}_f$ has a finite number of elements, all ends for the valuative tree.

So if $\nu \in \mathcal{V} \setminus \mathfrak{C}_f$, then $f_*\nu$ is a centered valuation, but not normalized generally. The norm will be $f_*\nu(\mathfrak{m}) = \nu(f^*\mathfrak{m})$: we can renormalize this valuation and obtain an action $f_\bullet : \mathcal{V} \setminus \mathfrak{C}_f \to \mathcal{V}$.

Definition 2.2.3. Let $f : (X, p) \to (Y, q)$ be a (dominant) holomorphic germ. For every valuation $\nu \in \mathcal{V}$ we define $c(f, \nu) := \nu(f^*\mathfrak{m})$ the **attraction rate of f along** ν; if $\nu = \nu_\mathfrak{m}$ is the multiplicity valuation, then we simply write $c(f) := c(f, \nu_\mathfrak{m})$ the **attraction rate** of f. For every valuation $\nu \in \mathcal{V} \setminus \mathfrak{C}_f$ we define $f_\bullet\nu := f_*\nu/c(f, \nu) \in \mathcal{V}$. If $f : (X, p) \to (X, p)$, we will also define $c_\infty(f) := \lim_{n\to\infty} \sqrt[n]{c(f^n)}$ the **asymptotic attraction rate** of f.

Up to fix coordinates in p and q, we can consider a germ $f : (X, p) \to (Y, q)$ as a germ $f : (\mathbb{C}^2, 0) \to (\mathbb{C}^2, 0)$: from now on we will state results in the latter case, but they can be easily extended to the general case.

Proposition 2.2.4 ([27, Proposition 2.4]). *The map $f_\bullet : \mathcal{V} \setminus \mathfrak{C}_f \to \mathcal{V}$ preserves the type of valuations (divisorial, irrational, curve, infinitely singular). Moreover, there exists an integer $N \geq 1$ such that for every $\nu \in \mathcal{V}$ the set $f_\bullet^{-1}\{\nu\}$ has cardinality at most N.*

We will need to know the geometric behavior of $f_\bullet$, especially when evaluated on divisorial or curve valuations.

Proposition 2.2.5 ([27, Proposition 2.5]). *Let ν be a divisorial valuation, and set $\nu' := f_\bullet\nu$. Then there exist modifications $\pi : X \to (\mathbb{C}^2, 0)$*

and $\pi' : X' \to (\mathbb{C}^2, 0)$*, and exceptional components* $E \in \Gamma^*_\pi$, $E' \in \Gamma^*_{\pi'}$ *such that* $\nu = \nu_E$, $\nu' = \nu_{E'}$ *and such that the map* f *lifts to a holomorphic map* $\hat{f} : X \to X'$ *sending* E *onto* E'*. Moreover,*

$$c(f, \nu_E) = \frac{b_{E'}}{b_E} k,$$

with b_E, b'_E *the generic multiplicities of* ν_E *and* $\nu_{E'}$ *respectively, and* $k \geq 1$ *is the largest integer such that* $\hat{f}^* E' \geq kE$ *as divisors.*

Proposition 2.2.6 ([27, Proposition 2.6]). *Let* C *be an analytic irreducible curve such that* $f(C) \neq \{0\}$ (i.e., $\nu_C \notin \mathfrak{C}_f$)*. Then* $C' := f(C)$ *is an analytic irreducible curve and* $f_\bullet \nu_C = \nu_{C'}$*. Further,*

$$c(f, \nu_C) = \frac{m(C')}{m(C)} e(f, C),$$

where $e(f, C) \in \mathbb{N}^*$ *denotes the topological degree of the restriction* $f|_C : C \to C'$*.*

Remark 2.2.7. The proof of Proposition 2.2.6 shows that a similar result is also valid for formal irreducible curves. In particular, if $t \mapsto \Phi(t)$ and $t \mapsto \Psi(t)$ are parametrizations of C and C' respectively, then the topological degree of $f|_C : C \to C'$ is defined by $e(f, C) = m(\Psi^{-1} \circ f \circ \Phi(t))$, where m denotes the multiplicity function on $\mathbb{C}[[t]]$.

In order to have an action on $\mathcal{V}$, we should extend $f_\bullet$ to contracted critical curve valuations.

Proposition 2.2.8 ([27, Proposition 2.7]). *Suppose* C *is an irreducible curve germ such that* $f(C) = \{0\}$ (i.e. $\nu_C \in \mathfrak{C}_f$)*. Then* $c(f, \nu_C) = \infty$*. Further, the limit of* $f_\bullet \nu$ *as* ν *increases to* ν_C *exists, and it is a divisorial valuation that we denote by* $f_\bullet \nu_C$*. It can be interpreted geometrically as follows. There exist modifications* $\pi : X \to (\mathbb{C}^2, 0)$ *and* $\pi' : X' \to (\mathbb{C}^2, 0)$*, such that* f *lifts to a holomorphic map* $\hat{f} : X \to X'$ *sending* C *to a curve germ included in an exceptional component* $E' \in \Gamma^*_{\pi'}$*, for which* $f_\bullet \nu_C = \nu_{E'}$*.*

For other properties of the action $f_\bullet$, we refer to [27, Section 2].

Regularity

Theorem 2.2.9 ([27, Theorem 3.1]). *Let* $f : (\mathbb{C}^2, 0) \to (\mathbb{C}^2, 0)$ *be a (dominant) holomorphic germ. Then* $f_\bullet : \mathcal{V} \to \mathcal{V}$ *is a regular tree map. Moreover any segment* I *where* $f_\bullet|_I$ *is an homeomorphism with the image can be decomposed into a finite number of segments where we have* $\alpha(f_\bullet \nu) = \frac{a + b\alpha(\nu)}{c + d\alpha(\nu)}$*, for some* $a, b, c, d \in \mathbb{N}$ *with* $ad \neq bc$*.*

The situation is much simpler when we deal with invertible germs.

Proposition 2.2.10. *Let $\Phi : (\mathbb{C}^2, 0) \to (\mathbb{C}^2, 0)$ be a (dominant) invertible holomorphic germ. Then the skewness α, the multiplicity m and the thinness A are invariant for the action of $\Phi_\bullet : \mathcal{V} \to \mathcal{V}$.*

Proof. Since Φ is invertible, the action $\Phi^* : R \to R$ is an isomorphism. It follows then directly from the definition of skewness that α is invariant for the action of $\Phi_\bullet$.

Analogously, the multiplicity of a curve is invariant up to biholomorphisms, and from the definition of multiplicity of a valuation as the minimum of the multiplicity of suitable curves, we get the invariance through the action of $\Phi_\bullet$ in this case.

Since the thinness is defined starting from skewness and multiplicity, the result follows. □

Remark 2.2.11. Let $f : (\mathbb{C}^2, 0) \to (\mathbb{C}^2, 0)$ be an holomorphic (dominant) germ, $\pi : X \to (\mathbb{C}^2, 0)$ and $\pi' : X' \to (\mathbb{C}^2, 0)$ be two modifications, and let $\hat{f} = (\pi')^{-1} \circ f \circ \pi : X \dashrightarrow X'$ be the lift of f. Then in general $\hat{f}$ is just defined as a rational map (being π and π' birational morphisms). Let us recall an useful criterion, based on the behavior of $f_\bullet$, for establishing if this rational map $\hat{f}$ is indeed holomorphic in a certain infinitely near point $p \in \pi^{-1}\{0\}$.

Proposition 2.2.12 ([27, Proposition 3.2]). *Let $\pi : X \to (\mathbb{C}^2, 0)$ and $\pi' : X' \to (\mathbb{C}^2, 0)$ be two modifications, and let $\hat{f} : X \dashrightarrow X'$ be the lift of f. For an infinitely near point $p \in \pi^{-1}\{0\} \subset X$, let $U(p) \subset \mathcal{V}$ be the open set of valuations whose center on X is p. Then $\hat{f}$ is holomorphic at p if and only if $f_\bullet U(p)$ does not contain any divisorial valuation associated to an exceptional component of π'. When $\hat{f}$ is holomorphic at p, the point $p' = \hat{f}(p)$ is characterized by $f_\bullet U(p) \subseteq U(p')$.*

Definition 2.2.13. Directly from Theorem 2.2.9, one can see that f induces a **tangent map** $d(f_\bullet)_\nu : T_\nu \mathcal{V} \to T_{f_\bullet \nu} \mathcal{V}$, for every $\nu \in \mathcal{V}$. Indeed, if $\nu \in \mathcal{V}$ is a valuation and $\overrightarrow{v} \in T_\nu \mathcal{V}$ is a tangent vector, then we can choose $\mu \in \overrightarrow{v}$ such that $f_\bullet$ is an homeomorphism between $[\nu, \mu]$ and $[f_\bullet \nu, f_\bullet \mu]$. So we can define $d(f_\bullet)_\nu(\overrightarrow{v})$ as the tangent vector in $f_\bullet \nu$ represented by $f_\bullet \mu$. It follows from the regularity of $f_\bullet$ that this is a good definition. We will often omit the point ν where we are considering the tangent map, and write $d(f_\bullet)_\nu = df_\bullet$.

The next proposition will show some geometrical properties that we can deduce from the behavior of the tangent map on a divisorial valuation.

Proposition 2.2.14 ([27, Proposition 3.3]). *Let $\nu \in \mathcal{V}$ be a divisorial valuation, and denote $\nu' = f_\bullet \nu$. Then let $\pi : X \to (\mathbb{C}^2, 0)$ and $\pi' : X' \to (\mathbb{C}^2, 0)$ be two modifications such that $\nu = \nu_E$ and $\nu' = \nu_{E'}$, with E, E' two exceptional components in Γ^*_π and $\Gamma^*_{\pi'}$ respectively. For every infinitely near point $p \in E$, let $\overrightarrow{v_p}$ be the tangent vector in ν associated to p, and analogously let $\overrightarrow{v_{p'}}$ be the tangent vector in ν' associated to $p' \in E'$. Then $df_\bullet(\overrightarrow{v_p}) = \overrightarrow{v_{p'}}$, with $p' = \hat{f}(p)$.*

Remark 2.2.15. When a divisorial valuation $\nu = \nu_\star$ is a fixed valuation for $f_\bullet$ (what we shall call eigenvaluation), applying Proposition 2.2.14, we have that $\hat{f}|_E : E \to E$ is a rational map from $E \cong \mathbb{P}^1(\mathbb{C})$ onto itself; so it admits a (non-critical) fixed point $p_\star$, and $df_\bullet$ admits a fixed tangent vector $\overrightarrow{v_{p_\star}}$ on $\nu_\star$.

Eigenvaluations and Basins of Attraction

In the previous sections we have recalled that $f_\bullet$ is a regular tree map (see Theorem 2.2.9), and that regular tree maps admit a fixed point (see Theorem 2.1.13): applying this result to $f_\bullet$ we obtain eigenvaluations.

Theorem 2.2.16 ([27, Theorem 4.2]). *Let $f : (\mathbb{C}^2, 0) \to (\mathbb{C}^2, 0)$ be a (dominant) holomorphic germ. Then there exists a valuation $\nu_\star \in \mathcal{V}$ such that $f_\bullet \nu_\star = \nu_\star$, and $c(f, \nu_\star) = c_\infty(f) =: c_\infty$. Moreover $\nu_\star$ cannot be a contracted critical curve valuation, and neither a non-analytic curve valuation if $c_\infty > 1$. If $\nu_\star$ is an end, then there exists $\nu_0 < \nu_\star$ (arbitrarily close to $\nu_\star$), such that $c(f, \nu_0) = c_\infty$, $f_\bullet$ preserves the order on $\{\nu \geq \nu_0\}$ and $f_\bullet \nu > \nu$ for every $\nu \in [\nu_0, \nu_\star)$. Finally, we can find $0 < \delta \leq 1$ such that $\delta c_\infty^n \leq c(f^n) \leq c_\infty^n$ for every $n \geq 1$.*

Definition 2.2.17. Let $f : (\mathbb{C}^2, 0) \to (\mathbb{C}^2, 0)$ be a (dominant) holomorphic germ. A valuation $\nu_\star \in \mathcal{V}$ is called **fixed valuation** for f if $f_\bullet \nu_\star = \nu_\star$. It is called **eigenvaluation** for f if it is a quasimonomial fixed valuation, or a fixed valuation which is a strongly attracting end.

Remark 2.2.18. In the rest of this chapter, we will always consider quasimonomial eigenvaluations whenever possible. Therefore when we will say that an eigenvaluation $\nu_\star$ is an end, we implicitly state that quasimonomial eigenvaluations do not exist.

Thanks to the next lemma, we can avoid a few cases for the type of an eigenvaluation.

Lemma 2.2.19 ([27, Lemma 7.7]). *For any $\nu \in \mathcal{V} \setminus \mathfrak{C}_f$ we have*

$$c(f, \nu) A(f_\bullet \nu) = A(\nu) + \nu(Jf),$$

where A denotes thinness and Jf is the Jacobian determinant of f.

Corollary 2.2.20. *Let f be a (dominant) holomorphic germ, and let $\nu_\star$ be an eigenvaluation for f. Then:*

(i) *if $c_\infty(f) > 1$ then $\nu_\star$ cannot be a non-analytic curve valuation;*
(ii) *if $c_\infty(f) = 1$ then $\nu_\star$ cannot be a quasimonomial valuation.*

Proof. The first assertion has been already stated in Theorem 2.2.16.

Let us suppose $c_\infty(f) = 1$. Then applying [27, Lemma 7.7] to the eigenvaluation (and recalling that $c_\infty(f) = c(f, \nu_\star)$ by Theorem 2.2.16) we obtain

$$A(\nu_\star) = A(\nu_\star) + \nu_\star(Jf),$$

that can be satisfied only if $A(\nu_\star) = \infty$. It follows that $\nu_\star$ cannot be a quasimonomial valuation. □

In order to study the behavior of a generic lift $\hat{f} : X \to X'$, we have to study the dynamics on the open sets $U(p)$ with p an infinitely near point.

Proposition 2.2.21 ([27, Proposition 5.2]). *Let f be a (dominant) holomorphic germ, and let $\nu_\star$ be an eigenvaluation for f.*

(i) *If $\nu_\star$ is an end for $\mathcal{V}$, then for any $\nu_0 \in \mathcal{V}$ with $\nu_0 \leq \nu_\star$, and ν_0 sufficiently close to $\nu_\star$, $f_\bullet$ maps the segment $I = [\nu_0, \nu_\star]$ strictly into itself and is order-preserving there. Moreover, if we set $U = U(\vec{v})$, where $\vec{v}$ is the tangent vector at ν_0 represented by $\nu_\star$, then $f_\bullet$ also maps the open set U strictly into itself and $f_\bullet^n \to \nu_\star$ as $n \to \infty$ in U.*
(ii) *If $\nu_\star$ is divisorial, then there exists a tangent vector $\vec{w}$ at $\nu_\star$ such that for any $\nu_0 \in \mathcal{V}$ representing $\vec{w}$ and sufficiently close to $\nu_\star$, $f_\bullet$ maps the segment $I = [\nu_\star, \nu_0]$ into itself and is order-preserving there. Moreover, if we set $U = U(\vec{v}) \cap U(\vec{w})$, where $\vec{v}$ is the tangent vector at ν_0 represented by $\nu_\star$, then $f_\bullet(I) \Subset I$, $f_\bullet(U) \Subset U$, and $f_\bullet^n \to \nu_\star$ as $n \to \infty$ on U.*
(iii) *If $\nu_\star$ is irrational, then there exist $\nu_1, \nu_2 \in \mathcal{V}$, arbitrarily close to $\nu_\star$, with $\nu_1 < \nu_\star < \nu_2$ such that $f_\bullet$ maps the segment $I = [\nu_1, \nu_2]$ into itself. Let $\vec{v_i}$ be the tangent vector at ν_i represented by $\nu_\star$ (for $i = 1, 2$), and set $U = U(\vec{v_1}) \cap U(\vec{v_2})$. Then $f_\bullet(U) \subseteq U$. Further, either $f_\bullet|_I^2 = \mathrm{id}_I$ or $f_\bullet^n \to \nu_\star$ as $n \to \infty$ on U.*

2.3. Rigidification

2.3.1. General result

Definition 2.3.1. Let $f : (\mathbb{C}^2, 0) \to (\mathbb{C}^2, 0)$ be a (dominant) holomorphic germ. Let $\pi : X \to (\mathbb{C}^2, 0)$ be a modification and $p \in \pi^{-1}(0)$

a point in the exceptional divisor of π. Then we shall call the triple $(\pi, p, \hat{f})$ a **rigidification** of f if the lift $\hat{f} = \pi^{-1} \circ f \circ \pi$ is a holomorphic rigid germ in p.

In this section we shall prove the existence of a rigidification for every germ. This is a generalization of a previous result of Favre and Jonsson (see [27, Theorem 5.1]). Here there are five cases instead of the four of [27, Theorem 5.1]: the new case is (ii), when we have a non-analytic curve eigenvaluation, and it arises only when we deal with f having a non-nilpotent differential. The other cases are treated as in [27, Theorem 5.1].

Theorem 2.3.2. *Every (dominant) holomorphic germ* $f : (\mathbb{C}^2, 0) \to (\mathbb{C}^2, 0)$ *admits a rigidification.*

Proof. Let $\nu_\star$ be an eigenvaluation for f (that exists thanks to Theorem 2.2.16). Let us split the proof into five cases, depending on the type of $\nu_\star$.

(i) Let $\nu_\star = \nu_C$ be a non-contracted analytic curve valuation. Pick ν_0 as in Proposition 2.2.21.(i). By increasing it, we can suppose ν_0 divisorial. We can even assume that there is a modification $\pi : X \to (\mathbb{C}^2, 0)$ such that $\nu_0 = \nu_{E_0}$ with $E_0 \in \Gamma_\pi^*$ an exceptional component, and such that $\tilde{C}$, the strict transform of C by π, has normal crossings (we can for example consider a modification π_1 such that $\nu_0 = \nu_{E_0}$, a modification π_2 such that the strict transform of C by π_2 has normal crossing, and then pick the join $\pi = \pi_1 \vee \pi_2$). Thanks to our choices for π there is a unique exceptional component $E \in \Gamma_\pi^*$ which intersects $\tilde{C}$ in a free point p. Since p is a free point, from Proposition 2.1.33 the best approximation of ν_C for π is unique, and this is ν_E. In particular $\nu_0 \leq \nu_E < \nu_C$, and ν_E can be chosen arbitrarily close to ν_C (by increasing ν_0). Moreover, the basin of attraction $U = U_{\nu_E}([\nu_\star])$ given by Proposition 2.2.21 is equal to $U(p)$ (see [26, Corollary 6.34]). From Proposition 2.2.21 it follows $f_\bullet U \Subset U$. From Proposition 2.2.12, the lift $\hat{f} = \pi^{-1} \circ f \circ \pi$ is holomorphic in p, and $\hat{f}(p) = p$.
By increasing ν_0, we may assume that U contains no critical curve valuations, or preimages of ν_C (by $f_\bullet$), except ν_C itself.
It follows that $\mathcal{C}^\infty(\hat{f}) = E$ or $\mathcal{C}^\infty(\hat{f}) = E \cup \tilde{C}$ has normal crossings. The first case arises when $\tilde{C}$ (the fixed curve) is not a critical curve for f (and we shall see that it can happen only when $c_\infty(f) = 1$). From Proposition 2.2.6, $\hat{f}(\tilde{C}) = \tilde{C}$, while E is contracted to p by $\hat{f}$ (since $f_\bullet \nu_E > \nu_E$); so $\mathcal{C}^\infty(\hat{f})$ is forward $\hat{f}$-invariant, and $\hat{f}$ is rigid.

(ii) Let $\nu_\star = \nu_C$ be a non-analytic curve valuation. Pick ν_0 as in Proposition 2.2.21. By increasing ν_0, we can suppose ν_0 divisorial. Let $\pi \in \mathcal{B}$ be a modification such that $\nu_0 = \nu_{E_0}$. From Proposition 2.1.33 there exists a unique best approximation ν_E of $\nu_\star$ for π (it is unique because $\nu_\star$ is an end of $\mathcal{V}$). We have $\nu_0 \leq \nu_E < \nu_C$, that can be chosen arbitrarily close to ν_C (by increasing ν_0). We consider now $U = U(p) = U_{\nu_E}([\nu_\star])$.
From Proposition 2.2.21 and Proposition 2.2.12, it follows $f_\bullet U \Subset U$, and the lift $\hat{f} = \pi^{-1} \circ f \circ \pi$ is holomorphic in p, and $\hat{f}(p) = p$. By shrinking $U(p)$, we can avoid all critical curve valuations.
It follows that $\mathcal{C}^\infty(\hat{f}) = E$ has normal crossings. Moreover, E is contracted to p by $\hat{f}$ (because $f_\bullet \nu_E > \nu_E$), $\mathcal{C}^\infty(\hat{f})$ is forward $\hat{f}$-invariant and $\hat{f}$ is rigid.

(iii) Let $\nu_\star$ be an infinitely singular valuation, and pick $\nu_0 < \nu_\star$ as in Proposition 2.2.21.(i) (we can assume ν_0 divisorial). The multiplicity function m is non-decreasing, integer-valued and unbounded on the segment $[\nu_{\mathfrak{m}}, \nu_\star)$, so there exist jump points for m arbitrarily close to $\nu_\star$. Let ν be such a point, with $\nu_0 \leq \nu < \nu_\star$. Recalling Proposition 2.1.25, these points are divisorial valuations, and from [26, Proposition 6.40] there exists a modification $\pi : X \to (\mathbb{C}^2, 0)$ such that $\nu = \nu_E$, with $E \in \Gamma_\pi^*$, and the center of $\nu_\star$ in X is a free point $p \in E$. Now we set $U = U(p)$: being $p \in \pi^{-1}(0)$, it follows that U contains no valuations ν_F for $F \in \Gamma_\pi^*$, and from Proposition 2.2.21 we have $f_\bullet U \Subset U$. As always, from Proposition 2.2.12 we have that the lift $\hat{f} = \pi^{-1} \circ f \circ \pi$ is holomorphic in p, and $\hat{f}(p) = p$; increasing ν_0 (*i.e.* shrinking $U(p)$), we can assume that U contains no critical curve valuations. So $\mathcal{C}^\infty(\hat{f}) = E$ has normal crossings. As in the previous case, $f_\bullet(\nu_E) > \nu_E$, so $\hat{f}(E) = p$, $\mathcal{C}^\infty(\hat{f})$ is forward $\hat{f}$-invariant and $\hat{f}$ is rigid.

(iv) Let $\nu_\star = \nu_E$ be a divisorial valuation, and pick ν_0 as in Proposition 2.2.21.(ii) (we can assume $\nu_0 = \nu_{E_0}$ divisorial). Pick a modification $\pi : X \to (\mathbb{C}^2, 0)$ such that $E, E_0 \in \Gamma_\pi^*$, and let $F \in \Gamma_\pi^*$ be the best approximation of E for π (see Proposition 2.1.33). Replace ν_0 by ν_F. The set U in Proposition 2.2.21 is of the form $U = U(p)$, with p the satellite point in $E \cap F$. As before $f_\bullet(U) \subseteq U$, and, by our choice of F and by Proposition 2.2.12, the lift $\hat{f} = \pi^{-1} \circ f \circ \pi$ is holomorphic in p, and $\hat{f}(p) = p$. By increasing ν_0 (*i.e.*, shrinking $U(p)$), we can assume that U contains no critical curve valuations for f. So $\mathcal{C}^\infty(\hat{f}) = E \cup F$ has normal crossings. Moreover, $f_\bullet^n \to \nu_\star$ on U, and so F is contracted to p by $\hat{f}$, while E is fixed; so $\mathcal{C}^\infty(\hat{f})$ is forward $\hat{f}$-invariant and $\hat{f}$ is rigid.

(v) Let $\nu_\star$ be an irrational valuation, and pick ν_1 and ν_2 such that $\nu_1 < \nu_\star < \nu_2$, as in Proposition 2.2.21.(iii), and let U be the basin of attraction determined by ν_1 and ν_2. Thanks to [27, Lemma 5.6], up to shrinking U, we can suppose that $U = U(p)$, with p an infinitely near point. From Proposition 2.2.12, we have that the lift $\hat{f} = \pi^{-1} \circ f \circ \pi$ is holomorphic in p, and $\hat{f}(p) = p$. Moreover, shrinking U, we can avoid critical curve valuations, and have $\mathcal{C}^\infty(\hat{f}) = E \cup F$ with normal crossings.
Let us study the dynamics on $\mathcal{C}^\infty(\hat{f})$. From Proposition 2.2.21, we have three cases: If $f_\bullet|_U = \mathrm{id}_U$, then both E and F are fixed by $f_\bullet$; if $f_\bullet|_U{}^2 = \mathrm{id}_U$ (but $f_\bullet|_U \neq \mathrm{id}_U$), then E and F are exchanged by $f_\bullet$; finally if $f_\bullet|_U{}^2 \neq \mathrm{id}_U$, then $f_\bullet U \Subset U$, and both E and F are contracted to p. In every case, we see that $\mathcal{C}^\infty(\hat{f})$ is forward $\hat{f}$-invariant and $\hat{f}$ is rigid. □

Remark 2.3.3. Studying the behavior of $\pi_\bullet$, with $\pi : (X, p) \to (\mathbb{C}^2, 0)$ a modification, we see that $\pi_\bullet$ is a bijection between $\mathcal{V}$ and $\overline{U(p)}$. Moreover, from the relation $\hat{f} = \pi^{-1} \circ f \circ \pi$, we see that $\pi_\bullet$ gives us a conjugation between $\hat{f}_\bullet$ and $f_\bullet|_{\overline{U(p)}}$. So from the dynamics of $f_\bullet$ on $\overline{U(p)}$, we can obtain informations on the rigidification $\hat{f}$. For example, when $f_\bullet^n \to \nu_\star$, then $\hat{f}$ will have a unique eigenvaluation $\pi_\bullet^{-1}\nu_\star$.

Remark 2.3.4. Let p be a satellite point, and E and F the exceptional components that contain p. Choose local coordinates (x, y) in p such that $E = \{x = 0\}$ and $F = \{y = 0\}$. Then considering $\pi_\bullet : \mathcal{V} \to \overline{U(p)}$, we have that $\pi_\bullet(\nu_x) = \nu_E$ and $\pi_\bullet(\nu_y) = \nu_F$. Moreover, $\pi_\bullet$ gives us a bijection between the segment $I = (\nu_x, \nu_y)$ of all (normalized) monomial valuations in (x, y) and (ν_E, ν_F). We can also parametrize I with valuations such as $\nu_{s,t} := \pi_* \mu_{s,t}$, with $\mu_{s,t}$ the monomial valuation on local coordinates (x, y) and of weights (s, t), satisfying $b_E s + b_F t = 1$. The latter condition is necessary to have $\nu_{s,t}$ normalized; b_E and b_F are the generalized multiplicity of ν_E and ν_F respectively (see [26, Ch. 6]).

2.3.2. Semi-superattracting case

In this section we deal with the semi-superattracting case, proving the uniqueness of the eigenvaluation in this case (see Theorem 2.3.8). We shall write

$$D_\lambda := \begin{pmatrix} \lambda & 0 \\ 0 & 0 \end{pmatrix}.$$

Lemma 2.3.5. *Let f be a (dominant) semi-superattracting holomorphic germ, such that $df_0 = D_\lambda$ with $\lambda \neq 0$, and let $\pi : X \to (\mathbb{C}^2, 0)$ be the* *single blow-up in* $0 \in \mathbb{C}^2$*, with* $E := \pi^{-1}(0) \cong \mathbb{P}^1(\mathbb{C})$ *the exceptional divisor. Set* $p = [1:0] \in E$*, and let* $\hat{f} : (X, p) \to (X, p)$ *be the lift of* f *through* π*. Then* $\hat{f}$ *is a semi-superattracting holomorphic germ, and* $d\hat{f}_p \cong D_\lambda$*.*

Proof. Since $df_0 = D_\lambda$, we have

$$f(x, y) = \big(\lambda x + f_1(x, y), f_2(x, y)\big), \tag{2.1}$$

with $f_1, f_2 \in \mathfrak{m}^2$. In the chart $\pi^{-1}(\{x \neq 0\})$ we can choose (u, t) coordinates in $p \in E$ such that

$$(x, y) = \pi(u, t) = (u, ut).$$

So for the lift $\hat{f} = \pi^{-1} \circ f \circ \pi$ we have

$$f \circ \pi(u, t) = \big(\lambda u + f_1(u, ut), f_2(u, ut)\big),$$

and then

$$\hat{f}(u, t) = \left(\lambda u + f_1(u, ut), \frac{f_2(u, ut)}{\lambda u + f_1(u, ut)}\right).$$

We have $u^2 \mid f_1(u, ut), f_2(u, ut)$; if we set $\hat{f} = (g_1, g_2)$, we have

$$g_1(u, t) = \lambda u\big(1 + O(u)\big)$$
$$g_2(u, t) = \frac{u^2 O(1)}{\lambda u\big(1 + O(u)\big)} = \alpha u + O(u^2),$$

with $\alpha = \lambda^{-1} a_{2,0}$, if $f_2(x, y) = \sum_{i+j \geq 2} a_{i,j} x^i y^j$. It follows that

$$d\hat{f}_p = \begin{pmatrix} \lambda & 0 \\ \alpha & 0 \end{pmatrix} \cong D_\lambda.$$

So $\hat{f}$ is a holomorphic germ with $d\hat{f}_p \cong D_\lambda$. □

Proposition 2.3.6. *Let* f *be a (dominant) semi-superattracting holomorphic germ such that* $df_0 = D_\lambda$ *with* $\lambda \neq 0$*,* $\nu_\star$ *an eigenvaluation for* f*, and* $(\pi, p, \hat{f})$ *a rigidification obtained from* $\nu_\star$ *as in Theorem 2.3.2. Then* $d\hat{f}_p \cong D_\lambda$ *and* $\nu_\star = \nu_C$ *is a (possibly formal) curve valuation, with* $m(C) = 1$*.*

Proof. For proving this result, we have to follow the proof of Theorem 2.1.13, as given in [27, Theorem 4.5], under the assumption $df_0 \cong D_\lambda$. Starting from any ν_0 (as in the proof of [27, Theorem 4.5]), we take any end $\nu_0' > f_\bullet \nu_0$, and we consider the induced tree map F_0 on $I_0 = [\nu_0, \nu_0']$. Let ν_1 be the (minimum) fixed point of F_0. Since $f_\bullet$ has no quasimonomial eigenvaluations (see Corollary 2.2.20), then $\nu_1 \geq f_\bullet \nu_0$. Up to choosing ν_0' such that $\nu_0' \notin d(f_\bullet)_{\nu_0}([f_\bullet \nu_0])$, we can suppose that $\nu_1 = f_\bullet \nu_0$.

Let us apply this argument for $\nu_0 = \nu_{\mathfrak{m}}$. If f is as in (2.1), then $\nu_1 = f_\bullet \nu_0$ is a divisorial valuation associated to an exceptional component E_1 obtained from the exceptional component E_0 of a single blow-up of $0 \in \mathbb{C}^2$ only by blowing-up free points (*i.e.*, the generic multiplicity $b(\nu_{E_1})$ of ν_{E_1} is equal to 1): as a matter of fact, $f_* \nu_0(x) = 1$ while $f_* \nu_0(\phi) \in \mathbb{N}$ for every $\phi \in R$.

Applying this argument recursively (as in the proof of [27, Theorem 4.5]), we get the assertion on the type of eigenvaluation.

For the result on $d\hat{f}_p$, we only have to observe that, on the proofs of Theorem 2.3.2 and [27, Theorem 5.1] in the case of an analytic curve eigenvaluation, up to shrink the basin of attraction, we can choose the infinitely near point p such that ν_{E_p} has generic multiplicity $b(\nu_{E_p})$ equal to 1, where E_p denotes the exceptional component obtained blowing-up p. Then, the modification π on the rigidification is the composition of blow-ups of free points, and then we can apply (recursively) Lemma 2.3.5 and obtain the thesis. □

Lemma 2.3.7. *Let f be a (dominant) semi-superattracting holomorphic germ such that $df_0 = D_\lambda$ with $\lambda \neq 0$. Then, up to a (possibly formal) change of coordinates, we can suppose that*

$$f(x, y) = \Big(\lambda x\big(1 + f_1(x, y)\big), y f_2(x, y)\Big),$$

with $f_1, f_2 \in \mathfrak{m}$.

Proof. First of all, we can suppose that

$$f(x, y) = \big(\lambda x + g_1(x, y), g_2(x, y)\big),$$

with $g_1, g_2 \in \mathfrak{m}^2$.

Thanks to Proposition 2.3.6, we know that there is an eigenvaluation $\nu_\star = \nu_C$ with $C = \{\phi = 0\}$ a (possibly formal) curve, with $\phi(x, y) = y - \theta(x)$ for a suitable θ. Up to the (possibly formal) change of coordinates $(x, y) \mapsto (x, y - \theta(x))$, we can suppose that $\phi = y$, and in particular, since C is fixed by f, that $y | g_2$. Then we have

$$f(x, y) = \Big(\lambda x\big(1 + f_1(x, y)\big) + h(y), y f_2(x, y)\Big),$$

with $f_1, f_2 \in \mathfrak{m}$ and $h \in \mathfrak{m}^2$. We shall denote $g_2(x, y) = yf_2(x, y)$.

Now we only have to show that up to a (possibly formal) change of coordinates, we can suppose $h \equiv 0$. We consider a change of coordinates of the form $\Phi(x, y) = (x + \eta(y), y)$, with $\eta \in \mathfrak{m}^2$. In this case we have $\Phi^{-1}(x, y) = (x - \eta(y), y)$. So we have

$$\begin{aligned} &\Phi^{-1} \circ f \circ \Phi(x, y) \\ &= \Big(\lambda\big(x + \eta(y)\big)\big(1 + f_1 \circ \Phi(x, y)\big) \\ &\qquad + h(y) - \eta \circ g_2 \circ \Phi(x, y), yf_2 \circ \Phi(x, y)\Big). \end{aligned} \tag{2.2}$$

We notice that the second coordinate of (2.2) is always divisible by y; we only have to show that there exists a suitable η such that the first coordinate of (2.2), valuated on $(0, y)$, is equal to 0. Hence we have to solve

$$\lambda\eta(y)\Big(1 + f_1\big(\eta(y), y\big)\Big) + h(y) - \eta \circ g_2\big(\eta(y), y\big) = 0. \tag{2.3}$$

If we set $\eta(y) = \sum_{n\geq 2} \eta_n y^n$, $h(y) = \sum_{n\geq 2} h_n y^n$, $1 + f_1(x, y) = \sum_{i+j\geq 0} f_{i,j} x^i y^j$ and $g_2(x, y) = \sum_{i+j\geq 2} g_{i,j} x^i y^j$ (with $g_{n,0} = 0$ for every n), then we have

$$\lambda \sum_{i+j\geq 0} f_{i,j} \sum_{H\in\mathbb{N}^{i+1}} \eta_H y^{|H|+j} + \sum_{n\geq 2} h_n y^n = \sum_k \eta_k \left(\sum_{i+j\geq 2} g_{i,j} \eta(y)^i y^j\right)^k. \tag{2.4}$$

Comparing the coefficients of y^n in both members, we get

$$\lambda\eta_n + \text{l.o.t.} = \text{l.o.t.},$$

where l.o.t. denotes a suitable function depending on η_h only for $h < n$. So thanks to (2.4) we have a recurrence relation for the coefficients η_n that is a solution of (2.3). □

Theorem 2.3.8. *Let f be a (dominant) semi-superattracting holomorphic germ. Then f admits a unique eigenvaluation $\nu_\star$, that has to be a (possibly formal) curve valuation with multiplicity $m(\nu_\star) = 1$. Let us denote $\nu_\star = \nu_C$, with $m(C) = 1$. Then, (only) one of the following holds:*

(i) *the set of valuations fixed by $f_\bullet$ consists only of the eigenvaluation $\nu_\star$, there exists (only) one contracted critical curve valuation ν_D, and in this case, it has to be $m(D) = 1$;*

(ii) *the set of valuations fixed by $f_\bullet$ consists of two valuations, the eigenvaluation $\nu_\star$, and a curve valuation ν_D, where D is a (possibly formal) curve with $m(D) = 1$.*

In both cases, C and D have transverse intersection, i.e., *their intersection number is* $C \cdot D = 1$.

Proof. Thanks to Lemma 2.3.7, we can suppose (up to formal conjugacy) that

$$f(x, y) = \Big(\lambda x\big(1 + g_1(x, y)\big), y g_2(x, y)\Big),$$

where $g_1, g_2 \in \mathfrak{m}$. We shall denote $f_2(x, y) = y g_2(x, y)$.

It follows that the eigenvaluation $\nu_\star$ given by Proposition 2.3.6 is $\nu_\star = \nu_w$, while ν_x is either fixed by $f_\bullet$ or a contracted critical curve valuation.

We only have to show that there are no other fixed valuations.

First of all, we want to notice what happens during the process used in the proof of Proposition 2.3.6 to tangent vectors at the valuation $\nu_0 = \nu_{\mathfrak{m}}$. Let us consider the family of valuations $\nu_{\theta,t}$, where $\theta \in \mathbb{P}^1(\mathbb{C})$ and $t \in [1, \infty]$, described as follows: if we denote $\phi_\theta = y - \theta x$ when $\theta \in \mathbb{C}$, and $\psi_\infty = x$, then $\nu_{\theta,t}$ is the valuation of skewness $\alpha(\nu_{\theta,t}) = t$ in the segment $[\nu_{\mathfrak{m}}, \nu_{\phi_\theta}]$, *i.e.*, the monomial valuation defined by $\nu_{\theta,t}(\phi_\theta) = t$ and $\nu_{\theta,t}(x) = 1$ if $\theta \in \mathbb{C}$, and $\nu_{\infty,t}(x) = t$, $\nu_{\infty,t}(y) = 1$.

Then we have that $\nu_1 = f_\bullet(\nu_{\mathfrak{m}}) = \nu_{0,m(f_2)}$, where m denotes the multiplicity function, while $f_\bullet(\nu_{\theta,t}) \geq \nu_1$ for every $\theta \in \mathbb{C}$ and t, since $f_\bullet(\nu_{\theta,t})(x) = 1 = \nu_1(x)$, $f_\bullet(\nu_{\theta,t})(y) \geq m(f_2) = \nu_1(y)$ and ν_1 is the minimum valuation that assumes those values on x and y.

We shall denote by $\overrightarrow{v_\theta}$ the tangent vector in $\nu_{\mathfrak{m}}$ represented by $\nu_{\theta,\infty}$, and by $\overrightarrow{u_\infty}$ the tangent vector in ν_1 represented by $\nu_{\mathfrak{m}}$; then it follows from what we have seen that $df_\bullet(\overrightarrow{v_\theta}) \neq \overrightarrow{u_\infty}$ for every $\theta \neq \infty$, and hence there are no fixed valuations in $\overline{U_{\nu_{\mathfrak{m}}}(\overrightarrow{v_\theta})}$ for every $\theta \neq 0, \infty$.

Moreover, applying this argument recursively as in the proof of Proposition 2.3.6, we obtain that there are no other fixed valuations in $\overline{U_{\nu_{\mathfrak{m}}}(\overrightarrow{v_0})}$, except for the eigenvaluation ν_w.

It remains to check for valuations in $\overline{U_{\nu_{\mathfrak{m}}}(\overrightarrow{v_\infty})}$. For this purpose, let us consider $f_\bullet(\nu_{\infty,t})$. For simplicity, we shall denote $\nu_{\infty,t} = \nu_{0,1/t}$ for every $t \in [0, 1]$ From direct computation, we have that $f_*(\nu_{\infty,t})(x) = t$, while

$$f_*(\nu_{\infty,t})(y) = \bigwedge_j (a_j t + b_j),$$

for suitable $a_j \in \mathbb{N}^*$, $b_j \in \mathbb{N}$. It follows that $f_\bullet(\nu_{\infty,t}) = \nu_{\infty,g(t)}$ for a suitable map $g(t)$, such that $g(t) < t$, and that $d(f_\bullet)_{\nu_{\infty,t}}([\nu_w]) = [\nu_w]$ (where the latter tangent vector belongs to the proper tangent space). Letting t go to ∞, we obtain that the only fixed valuation in $U_{\nu_{\mathfrak{m}}}(\overrightarrow{v_\infty})$ is ν_x, and we are done. □

Remark 2.3.9. Theorem 2.3.8 shows that every semi-superattracting germ f has two (formal) invariant curves: the first one C associated to the eigenvaluation, and hence to the eigenvalue λ of df_0; the second one D associated to the fixed or contracted critical curve valuation, and hence to the eigenvalue 0 of df_0. If f is of type $(0, \mathbb{C} \setminus \overline{\mathbb{D}})$, then both these curves are actually holomorphic, thanks to Theorem 1.1.33 (Stable-Unstable Manifold Theorem). In the general case of f of type $(0, \mathbb{C}^*)$, one can at least recover the manifold associated to the eigenvalue 0 of df_0, using Theorem 1.1.32 (Hadamard-Perron). In particular the curve D is always holomorphic. However C is not always holomorphic in general (see for example Proposition 2.7.5).

2.4. Rigid germs

In this section we will introduce the classification of attracting rigid germs in $(\mathbb{C}^2, 0)$ up to holomorphic and formal conjugacy (for proofs, see [25]), and the classification of rigid germs of type $(0, \mathbb{C} \setminus \mathbb{D})$ in $(\mathbb{C}^2, 0)$ up to formal conjugacy.

For stating them, we shall need 3 invariants.

- **The generalized critical set**: if $f : (\mathbb{C}^2, 0) \to (\mathbb{C}^2, 0)$ is a rigid germ then $C = \mathcal{C}^\infty(f)$ is a curve with normal crossings at the origin, and it can have 0, 1 or 2 irreducible components, that is to say that $\mathcal{C}^\infty(f)$ can be empty (if and only if f is a local biholomorphism in 0), an irreducible curve, or a reducible curve (with only 2 irreducible components); we will call f **regular**, **irreducible** or **reducible** respectively.
- **The trace**: if f is not regular, we have 2 cases: either $\operatorname{tr} df_0 \neq 0$, and df_0 has a zero eigenvalue and a non-zero eigenvalue, or $\operatorname{tr} df_0 = 0$, and df_0 is nilpotent.
- **The action on** $\pi_1(\Delta^2 \setminus \mathcal{C}^\infty(f))$: as $\mathcal{C}^\infty(f)$ is backward invariant, f induces a map from $U = \Delta^2 \setminus \mathcal{C}^\infty(f)$ (here Δ^2 denotes a sufficiently small polydisc) to itself, and so an action f_* on the first fundamental group of U. When f is irreducible, then $\pi_1(U) \cong \mathbb{Z}$, and f_* is completely described by $f_*(1) \in \mathbb{N}^*$ (f preserves orientation); when f is reducible, then $\pi_1(U) \cong \mathbb{Z} \oplus \mathbb{Z}$, and f_* is described by a 2×2 matrix with integer entries (in $\mathbb{N}$).

Definition 2.4.1. Let $f : (\mathbb{C}^2, 0) \to (\mathbb{C}^2, 0)$ a rigid germ. Then f belongs to:

Class 1 if f is regular;

Class 2 if f is irreducible, $\operatorname{tr} df_0 \neq 0$ and $f_*(1) = 1$;

Class 3 if f is irreducible, $\operatorname{tr} df_0 \neq 0$ and $f_*(1) \geq 2$;
Class 4 if f is irreducible, $\operatorname{tr} df_0 = 0$ (this implies $f_*(1) \geq 2$);
Class 5 if f is reducible, $\operatorname{tr} df_0 \neq 0$ (this implies $\det f_* \neq 0$);
Class 6 if f is reducible, $\operatorname{tr} df_0 = 0$ and $\det f_* \neq 0$;
Class 7 if f is reducible, $\operatorname{tr} df_0 = 0$ and $\det f_* = 0$.

Class	$\mathcal{C}^\infty(f)$	$\operatorname{tr} df_0$	$\det f_*$
1	0 (empty)		
2	1 (irreducible)	$\neq 0$	$= 1$
3			≥ 2
4		$= 0$	(≥ 2)
5	2 (reducible)	$\neq 0$	$(\neq 0)$
6		$= 0$	$\neq 0$
7			$= 0$

Remark 2.4.2. If f is irreducible, up to a change of coordinates we can assume $\mathcal{C}^\infty(f) = \{x = 0\}$. Then just using that $\{x = 0\}$ is backward invariant, we can write f in the form

$$f(x, y) = \Big(\alpha x^p\big(1 + \phi(x, y)\big), f_2(x, y)\Big),$$

with $\phi, f_2 \in \mathfrak{m}$. It can be easily seen that $f_* = p \geq 1$.

Analogously, if f is reducible, up to a change of coordinates we can assume $\mathcal{C}^\infty(f) = \{xy = 0\}$. Then just using that $\{xy = 0\}$ is backward invariant we can write f in the form

$$f(x, y) = \Big(\lambda_1 x^a y^b\big(1 + \phi_1(x, y)\big), \lambda_2 x^c y^d\big(1 + \phi_2(x, y)\big)\Big),$$

with $\phi_1, \phi_2 \in \mathfrak{m}$. In this case f_* is represented by the 2×2 matrix

$$M(f) := \begin{pmatrix} a & b \\ c & d \end{pmatrix}. \tag{2.5}$$

2.4.1. Attracting rigid germs

Now we will state some of the results proved in [25]; we recall normal forms only for classes that will be useful. We shall denote by $z\mathbb{C}[z]$ the set of all polynomials P such that $z \mid P$, and by $\mathfrak{m}_q[z] \subset z\mathbb{C}[z]$ the set of such polynomials whose degree is less than q.

Theorem 2.4.3 ([25, Ch.1]). *Let $f : (\mathbb{C}^2, 0) \to (\mathbb{C}^2, 0)$ be an attracting (holomorphic) rigid germ. Then f is locally holomorphically conjugated to one of the following:*

***Class* 1** *let α, β be the two non-zero eigenvalues for df_0, such as $(1 >) |\alpha| \geq |\beta| (> 0)$; if for any $k \in \mathbb{N}^*$ we have $\alpha^k \neq \beta$ (no resonance), then $f \cong (\alpha z, \beta w)$; if there exists $k \in \mathbb{N}^*$ such that $\alpha^k = \beta$ (resonance), then $f \cong (\alpha z, \beta w + \varepsilon z^k)$, with $\varepsilon \in \{0, 1\}$.*

***Class* 2** *$f \cong (\lambda z, wz^q + P(z))$, with $\lambda = \operatorname{tr} df_0$, $q \in \mathbb{N}^*$ and $P \in \mathfrak{m}_q[z]$.*

***Class* 3** *$f \cong (z^p, \lambda w)$, with $\lambda = \operatorname{tr} df_0$ and $p = f_*(1)$.*

***Class* 4** *$f \cong (z^p, \beta z^q w + Q(z))$, with $p = f_*(1)$, $\beta \neq 0$, $q \in \mathbb{N}^*$ and $Q \in z\mathbb{C}[z]$; more precisely, if $k := q/(p-1)$, then if $k \notin \mathbb{N}$ or $\beta \neq 1$ (non-special), then $Q \in \mathfrak{m}_q[z]$; otherwise if $k \in \mathbb{N}$ and $\beta = 1$ (special), then $Q(z) = P(z) + a_{k+q} z^{k+q}$ with $P \in \mathfrak{m}_q[z]$ and $a_{k+q} \in \mathbb{C}$.*

***Class* 5** *$f \cong (\lambda z, z^c w^d)$, with $\lambda = \operatorname{tr} df_0$, $c \geq 1$ and $d \geq 2$.*

***Class* 6** *$f \cong (\lambda_1 z^a w^b, \lambda_2 z^c w^d)$, with $\lambda_1\lambda_2 \neq 0$, $ad \neq bc$, $b + d \geq 2$, $a + c \geq 2$ or $(a, c) = (0, 1)$, and $ad = 0$ or $a + b, c + d \geq 2$; moreover, if $1 \notin \operatorname{Spec}(M(f))$ (if and only if $(a-1)(d-1) \neq bc$), we can suppose $\lambda_1 = \lambda_2 = 1$.*

***Class* 7** *$f \cong (\lambda_1 z^a w^b (1+\phi_1(z, w)), \lambda_2 z^c w^d (1+\phi_2(z, w)))$, with $\lambda_1\lambda_2 \neq 0$, $ad = bc$, and $\phi_1, \phi_2 \in \mathfrak{m}$.*

All germs in classes 2–4 have $C^\infty(f) = \{z = 0\}$, and all germs in classes 5–7 have $C^\infty(f) = \{zw = 0\}$.

The formal classification coincides with the holomorphic one, except for class 2, where one can suppose $P \equiv 0$.

Remark 2.4.4. During the proof of Theorem 2.4.3, in [25, Step 1 on page 491, and First case on page 498], the author starts from a germ of the form

$$f(z, w) = \Big(\alpha z^p\big(1 + g(z, w)\big), f_2(z, w)\Big),$$

with ϕ, $f_2 \in \mathfrak{m}$, and uses Theorem 1.1.7 (Kœnigs) and Theorem 1.1.8 (Böttcher) in [25, Theorem 3.1 and Theorem 3.2] respectively to assume, up to holomorphic conjugacy, that $g \equiv 0$ (and $\alpha = 1$ if $p \geq 2$). That argument does not work. Let us denote by $\Phi(z, w) = (\phi_w(z), w)$ the conjugation given by those theorems, and $\tilde{f} = \Phi \circ f \circ \Phi^{-1}$. We shall also denote $f(z, w) = (f_w^{(1)}(z), f_w^{(2)}(z))$, and analogously for $\tilde{f}$. By hypothesis $\phi_w(z)$ is such that $\phi_w \circ f_w^{(1)} \circ \phi_w^{-1}(z) = \alpha z^p$ (with $\alpha = 1$ if $p \geq 2$).

Then we have that

$$\tilde{f}_w^{(1)}(z) = \phi_{f_w^{(2)}(\phi_w^{-1}(z))} \circ f_w^{(1)} \circ \phi_w^{-1}(z),$$

and hence it does not coincide with αz^p.

We also note that if $|\alpha| > 1$, there still is Theorem 1.1.7 (Kœnigs), but the result is false (see Counterexample 2.4.11).

Nevertheless, one can obtain this result in the attracting case in the following way.

We want to solve the conjugacy relation

$$\Phi \circ f = e \circ \Phi, \tag{2.6}$$

where e is a germ of the form

$$e(z, w) = \big(\alpha z^p, e_2(z, w)\big),$$

with $e_2 \in \mathfrak{m}$.

We look for a solution of the form

$$\Phi(z, w) = \Big(z\big(1 + \phi(z, w)\big), w\Big),$$

with $\phi \in \mathfrak{m}$.

Then from the conjugacy relation (2.6) (comparing the first coordinate) we get

$$(1 + g)(1 + \phi \circ f) = (1 + \phi)^p.$$

Then we can consider

$$1 + \phi = \prod_{k=0}^{\infty} \left(1 + g \circ f^k\right)^{1/p^{k+1}}, \tag{2.7}$$

that would work if that product converges. But since f is attracting, there exists $0 < \varepsilon < 1$ such that $\|f(z, w)\| \leq \varepsilon \|(z, w)\|$, while since $g \in \mathfrak{m}$, there exists $M > 0$ such that $|g(z, w)| \leq M \|(z, w)\|$. It follows that

$$\sum_{k=0}^{\infty} p^{-(k+1)} \left|g \circ f^k(z, w)\right| \leq \sum_{k=0}^{\infty} \frac{M}{p} \left(\frac{\varepsilon}{p}\right)^k = \frac{M}{p - \varepsilon} < \infty,$$

and hence (2.7) defines an holomorphic germ ϕ, and hence a holomorphic map Φ that satisfies the conjugacy relation (2.6) in the first coordinate.

To choose e_2 such that (2.6) holds also for the second coordinate, we have to solve

$$f_2 = e_2 \circ \Phi,$$

but since Φ is a holomorphic invertible map, we can just define $e_2 = f_2 \circ \Phi$, and we are done.

Notice that this approach would not work for rigid germs of type $(0, \mathbb{C} \setminus \mathbb{D})$, not even formally.

2.4.2. Rigid germs of type $(0, \mathbb{C} \setminus \mathbb{D})$

Here we are going to study formal normal forms for rigid germs of type $(0, \mathbb{C} \setminus \mathbb{D})$. As notation, if $f(x, y) = \sum_{i,j} f_{i,j} x^i y^j$ is a formal power series, and $I = (i_1, \dots, i_k)$ and $J = (j_1, \dots, j_k)$ are two multi-indices, then we shall denote by $f_{I,J}$ the product

$$f_{I,J} = \prod_{l=1}^{k} f_{i_l, j_l}.$$

Moreover, when writing the dummy variables of a sum, we shall write the dimension of a multi-index after the multi-index itself. For example, $I(n)$ shall denote a multi-index $I \in \mathbb{N}^n$. We shall group together the multi-indices with the same dimension, separating these groups by a semi-colon, and we shall omit the dimension when it is equal to 1. For example,

$$\sum_{n,m; I, J(n); K(m)}$$

shall denote a sum over $n, m \in \mathbb{N}$, $I, J \in \mathbb{N}^n$ and $K \in \mathbb{N}^m$. As a convention, a multi-index of dimension 0 is an empty multi-index. We shall use a similar notation later for rigid germs in higher dimensions, see Subsection 3.1.1.

Remark 2.4.5. If $f : (\mathbb{C}^2, 0) \to (\mathbb{C}^2, 0)$ is a semi-superattracting holomorphic germ, recalling Remark 2.3.9, we have two invariant curves, C and D, with transverse intersection and multiplicity equal to 1, that play the role of the Unstable-Stable manifold (see 1.1.5). In particular, the formal conjugacy classes of $f|_C$ and $f|_D$ are formal invariants.

Moreover, up to formal conjugacy, we can suppose that $C = \{y = 0\}$ and $D = \{x = 0\}$. Let us set $f = (f_1, f_2)$; then, up to a formal change of coordinates, we can suppose that $f_1(x, 0)$ is equal to one of the formal normal forms given by Proposition 1.1.15.

Indeed, if $\phi \in \mathbb{C}[[x]]$ is the formal conjugation between $f_1(x, 0)$ and its formal conjugacy class $h(x)$, the formal map $\Phi(x, y) = (\phi(x), y)$ is a conjugation between f and a map g, with $g_1(\cdot, 0) = h(\cdot)$.

We shall refer at the normal form h of a germ f as the **first (formal) action** of f.

Lemma 2.4.6. *Let $f : (\mathbb{C}^2, 0) \to (\mathbb{C}^2, 0)$ be a semi-superattracting holomorphic germ. Then, up to formal conjugacy, we can suppose that*

$$f(x, y) = \big(h(x), g(x, y)\big),$$

with h the first action of f, and $g \in \mathfrak{m}^2$.

Proof. We can suppose that f is of the form

$$f(x, y) = \Big(\lambda x\big(1 + f_1(x, y)\big), g_2(x, y)\Big),$$

with $f_1 \in \mathfrak{m}$ and $y | g_2 \in \mathfrak{m}^2$.

We want to find a conjugation map of the form $\Phi(x, y) = (x(1 + \phi(x, y)), y)$ that conjugates f with

$$e(x, y) = \Big(\lambda x\big(1 + e_1(x)\big), e_2(x, y)\Big),$$

with $y|e_2 \in \mathfrak{m}^2$, and $\lambda x(1 + e_1(x)) = h(x)$.

Let us set $1+f_1(x, y) = \sum_{i+j\geq 0} f_{i,j} x^i y^j$, $g_2(x, y) = \sum_{i+j\geq 2} g_{i,j} x^i y^j$, $1 + \phi(x, y) = \sum_{i+j\geq 0} \phi_{i,j} x^i y^j$ and $1 + e_1(x) = \sum_{i\geq 0} e_i x^i$.

Then for the first coordinate of the conjugacy equation $\Phi \circ f = e \circ \Phi$ we have

$$\sum_{i+j\geq 0} \phi_{i,j} \lambda^{i+1} x^{i+1} \sum_{I,J\in\mathbb{N}^{i+1}} f_{I,J} x^{|I|} y^{|J|} \sum_{H,K\in\mathbb{N}^j} g_{H,K} x^{|H|} y^{|K|} \tag{2.8}$$

$$|| \tag{2.9}$$

$$\lambda \sum_h e_h x^{h+1} \sum_{N,M\in\mathbb{N}^{h+1}} \phi_{N,M} x^{|N|} y^{|M|}. \tag{2.10}$$

If we denote by $\mathbb{I}_{n,m}$ and by $\mathbb{III}_{n,m}$ the coefficients of $x^n y^m$ respectively of (2.8) and (2.10), we have

$$\mathbb{I}_{n,m} = \sum_{\substack{i,j;I,J(i+1);H,K(j)\\ i+1+|I|+|H|=n\\ |J|+|K|=m}} \phi_{i,j} \lambda^{i+1} f_{I,J} g_{H,K}; \qquad \mathbb{III}_{n,m} = \sum_{\substack{h;N,M(h+1)\\ h+1+|N|=n\\ |M|=m}} \lambda e_h \phi_{N,M}.$$

If we denote by lower order terms all terms depending on $\phi_{i,j}$ for (i, j) lower than the ones that appear in the equation (with respect to the lexicographic order), we get

$$\mathbf{1}_m^0 \phi_{n-1,0} \lambda^n (f_{0,0})^n + \text{l.o.t.} = \mathbb{I}_{n,m} = \mathbb{III}_{n,m} = \lambda e_0 \phi_{n-1,m} + \text{l.o.t.},$$

where **1** is the Kronecker's delta function. In particular, for $n = 0$ we have $0 = \mathbb{I}_{0,m} = \mathbb{II}_{0,m} = 0$ for every $m \in \mathbb{N}$, while for every $m \geq 1$ we have $\mathbb{I}_{n,m} = \text{l.o.t.}$ for every $n \in \mathbb{N}^*$. Since $\lambda e_0 = \lambda \neq 0$, we can use (2.9) to define recursively $\phi_{n,m}$ for every $m \geq 1$ once we have defined the base step for $m = 0$.

But the case $m = 0$ is exactly the same as consider the formal classification of $\tilde{f}(x) = \lambda x(1 + f_1(x, 0))$ as a map in one complex variable. Then, again recalling Remark 2.4.5 and putting all together, we can define a formal map Φ that solves the conjugacy relation $\Phi \circ f = e \circ \Phi$. □

We shall now give the formal classification of semi-superattracting rigid germs (the attracting case actually follows from Theorem 2.4.3).

Theorem 2.4.7. *Let $f : (\mathbb{C}^2, 0) \to (\mathbb{C}^2, 0)$ be a (holomorphic) semi-superattracting rigid germ. Let $\lambda \in \mathbb{C}^*$ be the non-zero eigenvalue of df_0.*

(i) *If $|\lambda| < 1$ or $\lambda = e^{2\pi i\theta}$ with $\theta \in \mathbb{R}\setminus\mathbb{Q}$, then f is formally conjugated to the map*

$$(z, w) \mapsto (\lambda z, z^c w^d).$$

(ii) *If $|\lambda| > 1$, then f is formally conjugated to the map*

$$(z, w) \mapsto \big(\lambda z, z^c w^d(1 + \varepsilon z^l)\big),$$

*where $\varepsilon \in \{0, 1\}$ if $\lambda^l = d$ (**resonant case**), and $\varepsilon = 0$ otherwise.*

(iii) *If there exists $r \in \mathbb{N}^*$ such that $\lambda^r = 1$, then f is formally conjugated to the map*

$$(z, w) \mapsto \big(\lambda z(1 + z^s + \beta z^{2s}), z^c w^d(1 + \varepsilon(z^r))\big),$$

where $r|s$ and $\beta \in \mathbb{C}$, while ε is a formal power series in z^r, and $\varepsilon \equiv 0$ if $d \geq 2$.

In all cases, $c \geq 0$, $d \geq 1$ and $c + d \geq 2$.

Proof. Thanks to Lemma 2.4.6 and simple considerations on rigid germs (see Remark 2.4.2), we can suppose that

$$f(x, y) = \Big(h(x), x^c y^d\big(1 + g(x, y)\big)\Big)$$

for a suitable $g \in \mathfrak{m}$, and where $h(x) = \lambda x(1 + \delta(x))$ is the first action of f.

We want to find a conjugation Ψ of the form $\Psi(x, y) = (x, y(1 + \psi(x, y)))$, between f and

$$e(x, y) = \Big(h(x), x^c y^d\big(1 + \varepsilon(x)\big)\Big),$$

for a suitable ε.

Then for the second coordinate of the conjugacy equation $\Psi \circ f = e \circ \Psi$ we have

$$x^c y^d \sum_{i+j\geq 0} \psi_{i,j}\lambda^i x^i \sum_{L\in\mathbb{N}^i} \delta_L x^{|L|} \sum_{I,J\in\mathbb{N}^{j+1}} g_{I,J} x^{|I|} y^{|J|} \tag{2.11}$$

$$|| \tag{2.12}$$

$$x^c y^d \sum_h \varepsilon_h x^h \sum_{H,K\in\mathbb{N}^d} \psi_{H,K} x^{|H|} y^{|K|}. \tag{2.13}$$

If we denote by $\mathbb{I}_{n,m}$ and by $\mathbb{III}_{n,m}$ the coefficients of $x^{c+n}y^{d+m}$ respectively of (2.11) and (2.13), we have

$$\mathbb{I}_{n,m} = \sum_{\substack{i,j;L(i);I,J(j+1)\\ i+cj+|L|+|I|=n\\ dj+|J|=m}} \psi_{i,j}\lambda^i \delta_L g_{I,J}; \qquad \mathbb{III}_{n,m} = \sum_{\substack{h;H,K(d)\\ h+|H|=n\\ |K|=m}} \lambda\varepsilon_h \psi_{H,K};$$

Then for $(n, m) \neq (0, 0)$ we get

$$\mathbf{1}_m^0 \psi_{n,0}\lambda^n + \text{l.o.t.} = \mathbb{I}_{n,m} = \mathbb{III}_{n,m} = d\psi_{n,m} + \text{l.o.t.},$$

where $\mathbf{1}$ is the Kronecker's delta function. Hence if $m > 0$, we can use (2.12) to define recursively $\psi_{n,m}$; for $m = 0$, we can have some resonance problems, when $\lambda^n = d$, that is exactly the condition expressed in (ii) and (iii). In these cases, studying the dependence of $\mathbb{III}_{n,0}$ on ε_h, we get

$$\mathbb{III}_{n,0} = \varepsilon_n + \text{l.o.t.},$$

where l.o.t. denotes here the dependence on lower order terms ε_h with $h < n$. So for each n that gives us a resonance, there exists a ε_n that satisfies $\mathbb{I}_{n,0} = \mathbb{III}_{n,0}$. Putting all together, and eventually performing a conjugacy by a linear map, we obtain the thesis. □

Remark 2.4.8. We notice that in the statement of Theorem 2.4.7, f belongs to Class 2 if and only if $d = 1$, to Class 3 if and only if $c = 0$, and to Class 5 otherwise.

Remark 2.4.9. The composition $\alpha \circ f_\bullet$, where α is either skewness or thinness, is not affected by slightly changing the non-null coefficients of a germ f (as far as we keep these coefficients non-null). What changes is the action of the differential $df_\bullet$ in suitable tangent spaces.

So the difference between normal forms in the resonant case of Theorem 2.4.7 lies in the action of $df_\bullet$, that is not invariant by change of coordinates, but has a very complicated behavior.

Remark 2.4.10. Let $\phi(x, y) = \sum \phi_{n,m} x^n y^m$ be a formal power series. Then ϕ is holomorphic (as a germ in 0) if and only if there is M such that

$$\left|\phi_{n,m}\right| \leq M\alpha^n \beta^m.$$

In particular, if ϕ is holomorphic, then $\limsup_n \sqrt[n]{\left|\phi_{n,m}\right|} < \infty$ for every $m \in \mathbb{N}$, and the same holds if we exchange the role of m and n.

We shall see now that when one has rigid germs of type $(0, \mathbb{C} \setminus \overline{\mathbb{D}})$, one cannot generally perform either the conjugacy of Lemma 2.4.6 or the one of Theorem 2.4.7 in a holomorphic way (this behavior is the opposite of the $(0, \mathbb{D})$ case).

Counterexample 2.4.11. Let us show that the conjugation given by Lemma 2.4.6 cannot be always holomorphic. Let $f(x, y) = (\lambda x(1 + y), xy)$ with $|\lambda| > 1$ and $e(x, y) = (\lambda x, e_2)$, and let

$$\Phi(x, y) = \Big(x\big(1 + \phi(x, y)\big), \psi(x, y)\Big)$$

be the (formal) conjugation given by Lemma 2.4.6. Let us use the same notations as in the proof of Theorem 2.4.7; then we already know that there is a formal conjugation Φ between f and e, with ϕ defined as in the proof of Lemma 2.4.6. In this case, it becomes:

$$\phi_{n,m} = \sum_{\substack{h,k,l \\ h+k=n \\ l+k=m}} \lambda^h \phi_{h,k} \binom{h+1}{l}. \tag{2.14}$$

Computing (2.14) for $m = 0$ we get

$$\phi_{n,0} = \lambda^n \phi_{n,0},$$

and hence $\phi_{n,0} = 0$ for every $n \neq 0$ (while we fixed $\phi_{0,0} = 1$). Computing (2.14) for $m = 1$ we get

$$\phi_{n,1} = (n+1)\lambda^n \phi_{n,0} + \lambda^{n-1} \phi_{n-1,1} = \mathbf{1}_n^0 + \lambda^{n-1} \phi_{n-1,1},$$

where $\mathbf{1}$ denotes the Kronecker's delta function, and hence

$$\phi_{n,1} = \lambda^{n(n-1)/2}.$$

Recalling Remark 2.4.10, we have that ϕ is not holomorphic.

Remark 2.4.12. Suppose that we have a germ $f = (\lambda x, f_2)$, with $|\lambda| > 1$, that is formally conjugated to $(\lambda x, x^c y^d)$ (*i.e.*, we are not in the resonance case). The proof of Theorem 2.4.7 shows also that the conjugation with the normal forms is unique when we ask it to be of the form $\Psi(x, y) = (x, y(1 + \psi))$. But if we consider a general conjugation map $\Phi = (\phi_1, \phi_2)$, since we have two invariant curves $D = \{x = 0\}$ and $C = \{y = 0\}$, then we have $x|\phi_1$ and $y|\phi_2$, and since the first coordinate is λx, from direct computation we also have that $\phi_1(x, y) = x$. So Φ is unique up to a linear change of coordinates. Hence, to prove that two germs are formally but not holomorphically conjugated, we only have to show that the conjugation found during the proof of Theorem 2.4.7 is not holomorphic.

Counterexample 2.4.13. Let us show that also the conjugation given by Theorem 2.4.7 cannot be always holomorphic. Let $f(x,y)=(\lambda x, xy(1+y))$ with $|\lambda| > 1$ and $e(x, y) = (\lambda x, xy)$, and let

$$\Psi(x, y) = \Big(x, w\big(1 + \psi(x, y)\big)\Big)$$

be the (formal) conjugation given by Theorem 2.4.7. Let us use the same notations as in the proof of Theorem 2.4.7; then we already know that there is a formal conjugation Φ between f and e, with ψ defined as in the proof of Theorem 2.4.7. In this case, it becomes:

$$\psi_{n,m} = \sum_{\substack{h,k,l \\ h+k=n \\ l+k=m}} \lambda^h \psi_{h,k} \binom{k+1}{l}. \tag{2.15}$$

Computing (2.15) for $m = 0$ we get

$$\psi_{n,0} = \lambda^n \psi_{n,0},$$

and hence $\psi_{n,0} = 0$ for every $n \neq 0$ (while we fixed $\psi_{0,0} = 1$). Computing (2.15) for $m = 1$ we get

$$\psi_{n,1} = \lambda^n \psi_{n,0} + \lambda^{n-1} \psi_{n-1,1} = \mathbf{1}_n^0 + \lambda^{n-1} \psi_{n-1,1},$$

where $\mathbf{1}$ denotes the Kronecker's delta function, and hence

$$\psi_{n,1} = \lambda^{n(n-1)/2},$$

and again we have that ψ is not holomorphic.

2.5. Formal classification of semi-superattracting germs

2.5.1. Invariants

In this subsection we shall introduce a few invariants for the formal classification of semi-superattracting germs, that arise from the rigidification process.

Remark 2.5.1. Let $f, e : (\mathbb{C}^2, 0) \to (\mathbb{C}^2, 0)$ be two (formal) germs and let us consider a conjugacy relation of the form

$$\begin{array}{ccc} (\mathbb{C}^2,0) & \xrightarrow{f} & (\mathbb{C}^2,0) \\ \downarrow{\scriptstyle \Phi} & & \downarrow{\scriptstyle \Phi} \\ (\mathbb{C}^2,0) & \xrightarrow{e} & (\mathbb{C}^2,0) \end{array} .$$

Then the induced action on the valuative trees

$$\begin{array}{ccc} \mathcal{V} & \xrightarrow{f_\bullet} & \mathcal{V} \\ \downarrow{\scriptstyle \Phi_\bullet} & & \downarrow{\scriptstyle \Phi_\bullet} \\ \mathcal{V} & \xrightarrow{e_\bullet} & \mathcal{V} \end{array} ,$$

gives a conjugacy relation with conjugation $\Phi_\bullet$.

Thanks to Proposition 2.2.10, we have that

$$\begin{aligned} \alpha\big(\Phi_\bullet(\nu)\big) &= \alpha(\nu), \\ m\big(\Phi_\bullet(\nu)\big) &= m(\nu), \\ A\big(\Phi_\bullet(\nu)\big) &= A(\nu), \end{aligned}$$

where α is the skewness, m is the multiplicity, and A is the thinness.

Hence for a germ $f : (\mathbb{C}^2, 0) \to (\mathbb{C}^2, 0)$, the action of $\alpha \circ f_\bullet$, $m \circ f_\bullet$, $A \circ f_\bullet$ are invariants by formal conjugacy.

Definition 2.5.2. Let $f : (\mathbb{C}^2, 0) \to (\mathbb{C}^2, 0)$ be a (dominant) holomorphic germ. We shall call **flexibility** of f the number

$$\operatorname{flex}(f) := \min \left\{ \operatorname{weight}(\pi) \mid (\pi, p, \hat{f}) \text{ is a rigidification for } f \right\}.$$

We shall say that f is r**-flexible** if $\operatorname{flex}(f) \leq r$.

Remark 2.5.3. The flexibility of a (dominant) germ f is well defined as a integer number $\mathrm{flex}(f) \in \mathbb{N}$ thanks to Theorem 2.3.2. Moreover, f is rigid if and only if $\mathrm{flex}(f) = 0$.

Finally, let us consider a conjugation Φ between two (dominant) germs f, e, and the induced action $\Phi_\bullet$ as in Remark 2.5.1.

A rigidification $(\pi, p, \hat{f})$ for f gives a basin of attraction $U(p) \subset \mathcal{V}$, while $\Phi_\bullet$ sends $U(p)$ in a suitable open set $U(q) \subset \mathcal{V}$.

Thanks to the invariance of parametrizations and multiplicity (see Proposition 2.2.10 and Remark 2.5.1), we have that q is an infinitely near point in the exceptional divisor of a suitable modification ρ such that $\mathrm{weight}(\pi) = \mathrm{weight}(\rho)$.

It follows that the flexibility of a germ is a formal invariant.

We shall focus now our attention on the semi-superattracting case.

Remark 2.5.4. Let $f : (\mathbb{C}^2, 0) \to (\mathbb{C}^2, 0)$ be a semi-superattracting (dominant) germ. Then thanks to Remark 2.4.5 and Lemma 2.4.6 we can suppose, up to formal conjugacy, that the two (formal) fixed curves C and D given by Theorem 2.3.8 (see also Remark 2.3.9) are defined by $C = \{y = 0\}$ and $D = \{x = 0\}$ respectively, and f is of the form

$$f(x, y) = \big(h(x), g(x, y)\big), \tag{2.16}$$

with h the first action of f, and $g \in \mathfrak{m}^2$.

We are now interested in simplifying the second coordinate of such a germ up to formal conjugacy, as we have done in Theorem 2.4.7 for the rigid case. We shall need some notations to make computations.

Definition 2.5.5. Let $f(x, y) = \big(h(x), g(x, y)\big)$ be a semi-superattracting rigid germ as in (2.16). Write $g(x, y) = \sum_{i,j} g_{i,j} x^i y^j$. For every $(i, j) \in \mathbb{N} \times \mathbb{N}$, we denote

$$\langle (i, j) \rangle = \{n, m \in \mathbb{N} \times \mathbb{N} \, : \, n \geq i \text{ and } m \geq j\}.$$

Then we set

$$\mathrm{W}(g) := \bigcup_{g_{i,j} \neq 0} \langle (i, j) \rangle,$$

and we denote by $\partial \mathrm{W}(g)$ a minimal set of generators of $\mathrm{W}(g)$, *i.e.*, a minimal subset of $\mathrm{W}(g)$ such that

$$\bigcup_{(i,j) \in \partial \mathrm{W}(g)} \langle (i, j) \rangle = \mathrm{W}(g).$$

We finally denote by $\mathrm{C}(\mathrm{W}(g))$ the set of vertices in the convex hull of $\mathrm{W}(g)$.

We shall not write explicitly the dependence on g if it is clear from the contest.

Remark 2.5.6. Let $f(x, y) = \big(h(x), g(x, y)\big)$ be a semi-superattracting rigid germ as in (2.16).

In Definition 2.5.5 we took $\partial \mathrm{W}g$ "a minimal set" of generators of $\mathrm{W}(g)$. As a matter of fact, $\partial \mathrm{W}(g)$ exists and it is unique.

If we set

$$\begin{aligned} p &:= \min\{i \ : \ (i, j) \in \mathrm{W}\}, \\ d_l &:= \min\{j \ : \ (i, j) \in \mathrm{W}, i \le l\} \qquad \text{(for } p \le l \le q\text{)}. \end{aligned}$$

then we have that the sequence (d_l) is decreasing, and hence it stabilizes, let us say at $l = q$:

$$d_{q-1} > d_q = d_{q+n} =: d \quad \text{for } n \in \mathbb{N},$$

where we denote $d_{p-1} = +\infty$ if $p = q$.

We shall also denote by $D = \{d_l\}$ the set of values of the sequence (d_l).

Then if for every $j \in D$ we set

$$c_j := \min\{i \mid d_i = j\},$$

then we can define $\partial \mathrm{W}(g)$ as

$$\partial \mathrm{W}(g) = \{(c_j, j) \mid j \in D\}.$$

With these definitions, we can write

$$f(x, y) = \left(h(x), \sum_{l=p}^{q} x^l y^{d_l} f^{(l)}\right), \tag{2.17}$$

where h is the first action, $f^{(l)} = f^{(l)}(y)$ for $p \le l < q$, and $f^{(q)} = f^{(q)}(x, y)$. Moreover, $f^{(l)}(0) \neq 0$ whenever $d_{l-1} > d_l$ (it includes $f^{(q)}(0, 0) \neq 0$).

Remark 2.5.7. Let $f : (\mathbb{C}^2, 0) \to (\mathbb{C}^2, 0)$ be a semi-superattracting (dominant) germ written as in (2.17). Let W be as in Definition 2.5.5, and we shall set q and d as in Remark 2.5.6.

Then f is r-flexible if and only if

$$\mathrm{W} \subseteq \{(i, j) \in \mathbb{N}^2 \mid r(j - d) + (x - q) \ge 0\}.$$

Indeed, in the proof of Theorem 2.1.13, to find an eigenvaluation as in Theorem 2.3.2, in our case we just have to consider blow-ups of free points, written in coordinates as $\pi(x, y) = (x, xy)$.

Moreover, thanks to Theorem 2.3.8, this is the only way to get an eigenvaluation.

Hence, we have just to lift f through π until we get a rigid germ.

By direct computation, we get the statement.

Remark 2.5.8. In the next proposition, starting from a semi-superattracting (dominant) germ $f : (\mathbb{C}^2, 0) \to (\mathbb{C}^2, 0)$ written as in (2.17), we shall consider the map $s : [0, +\infty] \to [0, +\infty]$ given by

$$s(t) := \bigwedge_{(i,j)\in \mathrm{W}} i + tj,$$

where W is defined as in Definition 2.5.5.

First of all, we notice that $(m, n) \in \langle (i, j) \rangle$ if and only if $m+tn \geq i+tj$ for every t, and it follows that

$$s(t) = \bigwedge_{(i,j)\in \partial \mathrm{W}} i + tj.$$

Next, if we consider two points $P_r = (i_r, j_r) \in \mathbb{N}^2$ for $r = 0, 1$, and we consider the segment between them, with $P_r = (i_r, j_r) := rP_1 + (1-r)P_0$ for $r \in [0, 1]$, then

$$\bigwedge_{r\in[0,1]} i_r + tj_r = \bigwedge_{r=0,1} i_r + tj_r,$$

and hence it follows that

$$s(t) = \bigwedge_{(i,j)\in \mathrm{C(W)}} i + tj.$$

So $s(t)$ is uniquely determined by C(W). We shall now see that also the converse is true.

In fact, a point $(i, j) \in \mathrm{W}$ is in C(W) if and only if there exists a open semi-plane $V = \{(x, y) \in \mathbb{R}^2 \mid a(x - i) + b(y - j) > 0\}$ for suitable $a, b > 0$, such that $V \supset \mathrm{W} \setminus \{(i, j)\}$.

But then for every $(x, y) \in V$ we have

$$x + y\left(\frac{b}{a}\right) > i + j\left(\frac{a}{b}\right),$$

and we are done.

Proposition 2.5.9. *Let f be a (dominant) semi-superattracting holomorphic germ as in* (2.16).

Then the set C(W) *is a formal conjugacy invariant for f.*

Proof. Let us consider the induced map $f_\bullet : \mathcal{V} \to \mathcal{V}$ on the valuative tree, and in particular the restriction of $f_\bullet$ on the interval $I = [\nu_x, \nu_y]$. Let us denote by α the skewness parametrization, by ν_t the unique valuation in

$[\nu_{\mathfrak{m}}, \nu_y]$ with skewness $\alpha(\nu_t) = t$ if $t \geq 1$, and the unique valuation in $[\nu_{\mathfrak{m}}, \nu_x]$ with skewness $\alpha(\nu_t) = 1/t$ if $t \leq 1$. Then $f_\bullet \nu_t = \nu_{s(t)}$ with

$$s(t) = \bigwedge_{(i,j)\in W} i + tj.$$

Thanks to Remark 2.5.1, the function $s(t)$ is a formal conjugacy invariant. But thanks to Remark 2.5.8, C(W) is uniquely determined by the map $s(t)$, and we are done. □

2.5.2. Classification

Definition 2.5.10. In the following theorem, we shall use a few notations for subsets of $\mathbb{N}^2$ (or $\mathbb{Z}^2$). Let $W \subseteq \mathbb{Z}^2, k, a, b \in \mathbb{Z}$ and $(i, j) \in \mathbb{Z}^2$. Then we shall denote

$$\begin{aligned} kW &= \{(kn, km) \in \mathbb{Z}^2 \mid (n, m) \in W\}, \\ \overline{W} &= \{(n, -m) \in \mathbb{Z}^2 \mid (n, m) \in W\}, \\ (i, j) + W &= \{(i + n, j + m) \in \mathbb{Z}^2 \mid (n, m) \in W\}, \\ W_a &= \{(n, m) \in W \mid n \geq a\}, \\ W^b &= \{(n, m) \in W \mid n \leq b\}. \end{aligned}$$

We notice that in particular $\langle (i, j) \rangle = (i, j) + \mathbb{N}^2$. When writing a subset of $\mathbb{Z}^2$ modified by some of these operations, we shall implicitly intersect the result with $\mathbb{N}^2$, obtaining a subset of $\mathbb{N}^2$.

Moreover, on the points in $\mathbb{N}^2$ we shall consider the following lexicographic order:

$$(n_1, m_1) > (n_2, m_2) \text{ iff } m_1 > m_2, \text{ or } m_1 = m_2 \text{ and } n_1 > n_2;$$

and we shall denote by l. o. t. terms of lower order with respect to this lexicographic order on the indices. In the case we are considering an equation that depends on several variables, we shall denote by l. o. t.(ϕ) the terms of lower order with respect to indices of the formal power series ϕ.

Theorem 2.5.11. *Let $f = (f_1, f_2)$ be a (dominant) semi-superattracting holomorphic germ. Let h be the first action, λ the non-zero eigenvalue of df_0, and $d = d_q$ as in Remark 2.5.6. Then, up to formal conjugacy, we have*

$$f(x, y) = \left(h(x), \sum_{l=p}^{q} x^l y^{d_l} f^{(l)} \right)$$

(with $f^{(l)}$ as in Remark 2.5.6), so that:

(i) *if $\lambda^u \neq d$ for every $u \in \mathbb{N}^*$* (**no resonance**)*, then $f^{(q)} \in \mathbb{C}^*$. This normal form is unique up to conjugations of the form $\Psi(x, y) = (\alpha_1 x, \alpha_2 y)$, with $\alpha_1, \alpha_2 \in \mathbb{C}^*$;*

(ii) *if there exists $u \in \mathbb{N}^*$ such that $\lambda^u = d \geq 2$* (**degenerate, simple resonance**)*, then $f^{(q)} = a_0 + a_u x^u$, with $a_0 \in \mathbb{C}^*$ and $a_u \in \mathbb{C}$. In this case, $h(x) = \lambda x$, and this normal form is unique up to conjugations of the form $\Psi(x, y) = (\alpha z, w(\psi_{0,0} + \psi_{u,0} x^u))$, with $\alpha, \psi_{0,0} \in \mathbb{C}^*$ and $\psi_{u,0} \in \mathbb{C}$;*

(iii) *if there exists $u \in \mathbb{N}^*$ such that $\lambda^u = d = 1$* (**non-degenerate, full resonance**)*, then $f^{(q)} = f^{(q)}(x^u)$, with $f^{(q)}(0) \in \mathbb{C}^*$. This normal form is unique up to conjugations of the form $\Psi(x, y) = (\alpha x, y\psi(x^u))$, with $\alpha, \psi(0) \in \mathbb{C}^*$.*

Proof. Recalling Remark 2.5.6, we can suppose that f is of the form

$$f(x, y) = \left(h(x), \sum_{l=p}^{q} x^l y^{d_l} f^{(l)}\right).$$

We will write $h(x) = \lambda x g(x)$, with $g(0) = 1$.

We shall call e the normal form candidate, of the form

$$e(x, y) = \left(h(x), \sum_{l=p}^{q} x^l y^{d_l} e^{(l)}\right),$$

with $e^{(l)} = e^{(l)}(y)$ for $p \leq l < q$ and $e^{(q)} = e^{(q)}(x)$. We shall look for a conjugation map Φ of the form $\Phi(x, y) = (x, y(1 + \psi))$: since we have to fix the coordinate axes in order to maintain the special form and thanks to Remark 2.4.12 and some considerations that arise from the proof of Proposition 1.1.15, we have no loss of generality considering only this kind of conjugation map; so we want to solve the conjugacy equation $\Phi \circ f = e \circ \Phi$.

We shall set $g(x) = \sum_h g_h x^h$; moreover $f^{(l)}(y) = \sum_k f_k^{(l)} y^k$ for $p \leq l < q$, $f^{(q)}(x, y) = \sum_{i,j} f_{i,j}^{(q)} x^i y^j$, $e^{(l)}(y) = \sum_k e_k^{(l)} y^k$ for $p \leq l < q$, $e^{(q)}(x) = \sum_h e_h^{(q)} x^h$ and $1 + \psi(x, y) = \sum_{i,j} \psi_{i,j} x^i y^j$. Sometimes it shall be useful to denote some coefficients that depend only by one parameter, as coefficients that depend on two parameters: for example, we shall write $f_j^{(l)} = f_{0,j}^{(l)}$.

Writing explicitly in formal power series the second coordinate of the conjugacy equation, we get

$$\sum_{i,j} \psi_{i,j}\lambda^i x^i \sum_{M(i)} g_M x^{|M|} \sum_{I,J,L(j+1)} x^{|L|} y^{|d_L|} f_{I,J}^{(L)} x^{|I|} y^{|J|} \tag{2.18}$$

$$|| \tag{2.19}$$

$$\sum_{l=p}^{q-1} x^l \sum_k e_k^{(l)} y^{d_l+k} \sum_{H,K(d_l+k)} \psi_{H,K} x^{|H|} y^{|K|} + y^d \sum_h e_h^{(q)} x^{q+h} \sum_{H,K(d)} \psi_{H,K} x^{|H|} y^{|K|}. \tag{2.20}$$

If we denote by $\mathbb{I}_{n,m}$ and by $\mathbb{II}_{n,m}$ the coefficients of $x^n y^m$ respectively of (2.8) and (2.10), we have

$$\mathbb{I}_{n,m} = \sum_{\substack{i,j;M(i),I,J,L(j+1)\\ i+|M|+|L|+|I|=n\\ |d_L|+|J|=m}} \psi_{i,j}\lambda^i g_M f_{I,J}^{(L)}; \quad \mathbb{II}_{n,m} = \sum_{\substack{l,k;H,K(d_l+k)\\ l+|H|=n\\ d_l+k+|K|=m}} e_k^{(l)} \psi_{H,K} + \sum_{\substack{h;H,K(d)\\ q+h+|H|=n\\ d+|K|=m}} e_h^{(q)} \psi_{H,K}; \tag{2.21}$$

we shall also denote $\mathbb{E}_{n,m} = \mathbb{II}_{n,m} - \mathbb{I}_{n,m}$.

So we have to solve the equations

$$\mathbb{E}_{n,m} = 0 \tag{2.22}$$

for every $(n, m) \in \mathbb{N} \times \mathbb{N}$, with respect to ψ, while we would like to set as many coefficients of e as possible equal to 0.

If $(n, m) \notin \mathrm{W}$, then $\mathbb{E}_{n,m} = 0$, so equation (2.22) is automatically satisfied.

Let us consider $(n, m) \in \mathrm{W}$. Then we have that

$$\mathbb{E}_{n,m} = \widetilde{R}_{n,m}\left(\begin{array}{c|l} \psi_{i,j} & (i, j) \in (n, m) - \mathrm{W}, \\ e_h^{(l)} & (l, h) \in (0, m) + \overline{\mathrm{W}}^{n\wedge q} \end{array} \right),$$

for a suitable function $\widetilde{R}_{n,m}$, where we have explicited the dependence from e and ψ.

First of all, we notice that

$$\mathbb{E}_{q,d} = e_0^{(q)} \psi_{0,0}^d - f_{0,0}^{(q)} \psi_{0,0},$$

and, since $\psi_{0,0} = 1$, from equation (2.22) applied to (q, d) we get that $e_0^{(q)} = f_0^{(q)}$ is an invariant.

Let us consider $(n, m) \in \langle (q, d) \rangle \setminus \{(q, d)\} \subset \mathrm{W}$. Then we have

$$\mathbb{E}_{n,m} = d e_0^{(q)} \psi_{n-q,m-d_q} - \delta_m^q \lambda^{n-q} \psi_{n-q,0} f_{0,0}^{(q)} + \text{l.o.t.}(\psi). \tag{2.23}$$

We have already seen that $e_0^{(q)} = f_{0,0}^{(q)}$, so the leading coefficient is reduced to $c_{n-q,m-d} := e_0^{(q)}(d - \delta_m^q \lambda^{n-q})$. Whenever $c_{n,m} \neq 0$, from $\mathbb{E}_{q+n,d+m}$ we can define recursively

$$\psi_{n,m} = R_{n,m}\big(e_h^{(l)} \mid (l,h) \in (0, d_q + m) + \overline{\mathrm{W}}^q\big), \tag{2.24}$$

with $R_{n,m}$ a suitable function, depending on the indicated coefficients $e_h^{(l)}$ and the other given datas (such as f and λ).

This is always the case if we are in case (i), while $c_{u,0} = 0$ if we are in case (ii), and $c_{ku,0} = 0$ for every $k \in \mathbb{N}^*$ if we are in case (iii).

In all these cases, the leading term of $\mathbb{E}_{q+n,d}$ with respect to $e^{(q)}$ is

$$\mathbb{E}_{q+n,d} = e_n^{(q)} + \text{l.o.t.}(e^{(q)}). \tag{2.25}$$

Since the leading term $\mathbb{E}_{q+n,d}$ with respect to ψ is vanishing thanks to resonances, we can define (recursively in case (iii)) $e^{(q)}$. We notice that the definition of $e^{(q)}$ is not unique in case (iii), since it depends on the choice we have on $\psi_{ku,0}$ for $k \in \mathbb{N}^*$, while it is uniquely determined in case (ii) (and obviously (i)).

Consider now $(n,m) \in \mathrm{W} \setminus \langle (q,d) \rangle$. In this case we have

$$\begin{aligned}\mathbb{E}_{n,m} &= e_{m-d_n}^{(n)} + \text{l.o.t.}(e)\\ &= \widetilde{R}_{n,m}\big(\psi_{i,j} \mid (i,j) \in (n,m) - \mathrm{W};\, e_h^{(l)} \mid (l,h) \in (0,m) + \overline{\mathrm{W}}^n\big).\end{aligned}$$

We want to show, using equation (2.24) for explicitating the dependence of $\psi_{i,j}$ from $e_h^{(l)}$, that actually the $\phi_{i,j}$'s with $(i,j) \in (n,m) - \mathrm{W}$ depend only on coefficients of e smaller than $e_{m-d_n}^{(n)}$ (with respect to the lexicographic order previously defined). Indeed, the maximum j that appears is $j = m - d_n$, and in this case $i \leq n-1$ and we are done. So from $\mathbb{E}_{d_l+h}^{(l)}$ for $p \leq l < q$ and $h \geq 0$ we can define recursively

$$e_h^{(l)} = S_h^{(l)}\big(e_k^g \mid (g,k) \in (0, d_l + h) + \overline{\mathrm{W}}^l\big).$$

Hence we have found e, and a Φ that solves the conjugacy equation, and we are done. □

2.6. Rigid germs of type (0, 1)

Definition 2.6.1. Let $f : (\mathbb{C}^2, 0) \to (\mathbb{C}^2, 0)$ be a holomorphic germ of type $(0, 1)$. Thanks to Theorem 2.3.8, we have a (formal) curve $C = \{y = \theta(x)\}$ that is invariant for the action of f. Moreover, $f|_C$ is a (formal) tangent to the identity map from C to C, *i.e.*, with $\big(f|_C\big)'(0) = 1$. We shall call **parabolic multiplicity** of f the parabolic multiplicity of $f|_C$.

Remark 2.6.2. The definition of parabolic multiplicity just given coincide with the definition of multiplicity in [33].

Definition 2.6.3. Let R and ρ be two positive real numbers, and $k \in \mathbb{N}^*$. Then we define

$$\begin{aligned} P_R &:= \{x \in \mathbb{C} \mid \operatorname{Re} x > R\}, \\ V_{R,\rho} &:= \{(x, y) \in \mathbb{C}^2 \mid x \in P_R, |y| < \rho\}. \end{aligned}$$

Let us denote by D_R and $U_{R,\rho}$ the images of P_R and $V_{R,\rho}$ under the inversion in the first coordinate $(x, y) \mapsto (x^{-1}, y)$, so we have

$$\begin{aligned} D_R &:= \left\{x \in \mathbb{C} \mid \left|x - \frac{1}{2R}\right| < \frac{1}{2R}\right\}, \\ U_{R,\rho} &:= \{(x, y) \in \mathbb{C}^2 \mid x \in D_R, |y| < \rho\}. \end{aligned}$$

There are k branches of $\sqrt[k]{x}$ in D_R (since $0 \notin D_R$). Let $\{\Delta_{R,j}\}_{j=0,\dots,k-1}$ be the images of D_R by these determinations. Let us define

$$W_{R,\rho,j} := \{(x, y) \in \mathbb{C}^2 \mid x \in \Delta_{R,j}, |y| < \rho\}, \quad \text{for } j = 0, \dots, k-1.$$

Theorem 2.6.4 ([33, Section 4]). *Let $f : (\mathbb{C}^2, 0) \to (\mathbb{C}^2, 0)$ be a holomorphic germ of type $(0, 1)$, with parabolic multiplicity $k + 1$. Then for every $j = 0, \dots, k-1$, and for suitable $R, \rho > 0$ small enough, there exists holomorphic maps*

$$\phi_j : W_{R,\rho,j} \to \mathbb{C}$$

such that

$$\phi\big(f(p)\big) = \phi(p) + 1.$$

Moreover

$$\phi_j(x, y) = \frac{1}{x^k}\big(1 + O(x^k \log x, y)\big).$$

Remark 2.6.5. Recalling Remark 2.4.2, if we have a holomorphic rigid germ of type $(0, 1)$, then we can suppose that

$$f(x, y) = \Big(x\big(1 + g(x, y)\big), x^c y^d\big(1 + h(x, y)\big)\Big), \tag{2.26}$$

where $g(0, 0) = h(0, 0) = 0$.

Lemma 2.6.6. *Let $f : (\mathbb{C}^2, 0) \to (\mathbb{C}^2, 0)$ be a holomorphic rigid germ of type $(0, 1)$ of parabolic multiplicity $k+1$. Then for every $j = 0, \dots, k-1$ and for suitable $R, \rho > 0$ small enough, if we set $W_j = W_{R,\rho,j}$, there* *exists a holomorphic conjugation* $\Phi_j : W_j \to \widetilde{W}_j$ *between* $f|_{W_j}$ *and the map*

$$(x, y) \mapsto \left(\frac{x}{\sqrt[k]{1 + x^k}}, x^c y^d \big(1 + h(x, y)\big) \right), \tag{2.27}$$

where $h(x, y) = O(x^k \log x, x, y)$, *and* $\widetilde{W}_j$ *is a suitable parabolic domain.*

Proof. Up to linear transformations, we can suppose that $j = 0$. Let us set

$$\Psi(x, y) = \left(\frac{1}{x^k}, y \right),$$
$$\Phi(x, y) = \big(\phi(x, y), y\big),$$

where $\phi = \phi_0$ is given by Theorem 2.6.4. If we consider the conjugation $\Upsilon := \Psi^{-1} \circ \Phi$, then we have that

$$\Upsilon(x, y) = \Psi^{-1}\Big(x^{-k}\big(1 + O(x^k \log x, y)\big), y\Big) = \Big(x\big(1 + O(x^k \log x, y)\big), y\Big).$$

By directly computing $\Upsilon \circ f \circ \Upsilon^{-1}$ we get the statement. □

Corollary 2.6.7. *Let* $f : (\mathbb{C}^2, 0) \to (\mathbb{C}^2, 0)$ *be a holomorphic rigid germ of type* $(0, 1)$ *of parabolic multiplicity* $k + 1$. *Let us set* $W = W_{R,\rho,j}$ *for suitable* $j \in \{0, \ldots, k - 1\}$, $R, \rho > 0$ *small enough. If there exists a parabolic curve* $\{y = \theta(x)\}$ *in* W, *then there exists a holomorphic conjugation* $\Phi : W \to \widetilde{W}$ *between* $f|_W$ *and the map*

$$(x, y) \mapsto \left(\frac{x}{\sqrt[k]{1 + x^k}}, x^c y^d \big(1 + h(x, y)\big) \right), \tag{2.27}$$

where $h(x, y) = O(x^k \log x, x, y)$, $d \geq 1$, *and* $\widetilde{W}$ *is a suitable parabolic domain.*

Proof. Thanks to Lemma 2.6.6, we just have to prove that, with the existence of a parabolic curve, we can get $d \geq 1$ in equation 2.27. Let us suppose that we can obtain (2.27) with $d = 0$. Then we just need to conjugate by $(x, y) \mapsto (x, y - \theta(x))$ and we are done. □

Theorem 2.6.8. *Let* $f : (\mathbb{C}^2, 0) \to (\mathbb{C}^2, 0)$ *be a holomorphic rigid germ of type* $(0, 1)$ *of parabolic multiplicity* $k + 1$ *that belongs to Class* 3 *or* 5 *(*i.e., $d \geq 2$ *if* f *is written as in* (2.26)*). Then for every* $j = 0, \ldots, k - 1$ *and for suitable* $R, \rho > 0$ *small enough, if we set* $W_j = W_{R,\rho,j}$, *there* *exists a holomorphic conjugation $\Phi_j : W_j \to \widetilde{W}_j$, with $\widetilde{W}_j$ a suitable parabolic domain, between $f|_{W_j}$ and the map*

$$\widetilde{f}(x, y) = \left(\frac{x}{\sqrt[k]{1 + x^k}}, x^c y^d \right); \tag{2.28}$$

equivalently, conjugating by $\Psi(x, y) = (x^{-k}, y)$, $f|_{W_j}$ is holomorphically conjugated to

$$(x, y) \mapsto (x + 1, x^{-c/k} y^d).$$

Proof. As always, we can suppose $j = 0$ and $W = W_0$. Thanks to Lemma 2.6.6, we can suppose that $f|_W$ is as in (2.27). Let us consider the maps

$$\Phi_n(x, y) = \big(x, y(1 + \phi_n(x, y)\big), \quad \text{with } 1 + \phi_n = \prod_{i=1}^{n} (1 + h \circ f^{i-1})^{d^{-i}}. \tag{2.29}$$

Then we have that

$$\begin{aligned} \big(\Phi_n \circ f\big)_2(x, y) &= x^c y^d \big(1 + h(x, y)\big) \prod_{i=1}^{n} \big(1 + h \circ f^i(x, y)\big)^{d^{-i}} \\ &= x^c y^d \prod_{i=1}^{n+1} \big(1 + h \circ f^{i-1}(x, y)\big)^{d^{1-i}} = x^c y^d \phi_{n+1}(x, y) \\ &= \big(\widetilde{f} \circ \Phi_{n+1}\big)_2(x, y). \end{aligned}$$

Hence if ϕ_n tends to a holomorphic map $\phi_\infty =: \phi$ in W, we are done.

But from (2.29), taking the log of the absolute value, we just have to prove that

$$\sum_{n=1}^{\infty} d^{-n} \left| h \circ f^{\circ n-1} \right|$$

converges in W. Since $h(x, y) = O(x^k \log x, x, y)$, for every $\varepsilon > 0$ there exists $M > 0$ big enough such that $|h(x, y)| \le M \left|(x, y)\right|^{1-\varepsilon}$, while since we are in W that is invariant for f, we have that there exists $C > 0$ such that $|f^n(x, y)| < C$. Then we have

$$\sum_{n=1}^{\infty} d^{-n} \left| h \circ f^{\circ n-1} \right| \le 2M \sum_{n=1}^{\infty} d^{-n} M C^{1-\varepsilon},$$

that converges if $d \ge 2$. □

Remark 2.6.9. Actually in the last estimate of the proof of Theorem 2.6.8, we just need that $|f^n(x, y)| \leq C\Lambda^n$ for a suitable $C > 0$, $0 < \Lambda < d$. In particular, if we have a germ of type $(0, \lambda)$ with $|\lambda| < d$ and such that the first coordinate depends only on x, then a result analogous to Theorem 2.6.8 holds. We notice that when $|\lambda| < d$, we have seen in the last section that we have no resonances, while exactly for $|\lambda| = d$ we can have them (in particular, when $\lambda = d$).

Lemma 2.6.10. *Let $f : (\mathbb{C}^2, 0) \to (\mathbb{C}^2, 0)$ be a holomorphic rigid germ of type $(0, 1)$ of parabolic multiplicity $k + 1$ as in (2.27), with $d = 1$ (i.e., it belongs to class 2). Set $(x_n, y_n) = f^{\circ n}(x_0, y_0)$. Then there exist $R, \rho > 0$ such that for every $j = 0, \dots, k - 1$ we have*

$$|x_n|^k = \frac{|x_0|^k}{\left|1 + nx_0^k\right|},$$
$$|y_n| \leq \left(2\,|x_0|^c\right)^n |y_0|,$$

in $W_{R,\rho,j}$. In particular,

$$|x_n| = O\left(\frac{1}{n^{1/k}}\right).$$

Proof. The proof is straightforward; we just notice that for the first coordinate of f^n, we have that it is conjugated through $(x, y) \mapsto (x^{-k}, y)$ to the map $x \mapsto x + n$, and hence it is of the form

$$x \mapsto \frac{x}{\sqrt[k]{1 + nx^k}}.$$

□

Theorem 2.6.11. *Let $f : (\mathbb{C}^2, 0) \to (\mathbb{C}^2, 0)$ be a holomorphic rigid germ of type $(0, 1)$ of parabolic multiplicity $k + 1$ that belongs to Class 2 (i.e., $d = 1$ if f is written as in (2.26), recalling Corollary 2.6.7). Let us set $W = W_{R,\rho,j}$ for suitable $j \in \{0, \dots, k - 1\}$, $R, \rho > 0$ small enough. If there exists a parabolic curve $\{y = \theta(x)\}$ in W, then there exists a holomorphic conjugation $\Phi : W \to \widetilde{W}$ between $f|_W$ and the map*

$$\widetilde{f}(x, y) = \left(\frac{x}{\sqrt[k]{1 + x^k}}, x^c y\big(1 + \widetilde{h}(x)\big)\right), \tag{2.30}$$

with $\widetilde{h}(x) = O(x^k \log x, x)$.

Proof. Thanks to Lemma 2.6.6, we can suppose that $f|_W$ is as in (2.27). We shall write it on the form

$$(x, y) \mapsto \big(\widetilde{g}(x), x^c y^d \big(1 + \widetilde{h}(x) + k(x, y)\big)\big). \tag{2.27}$$

Let us consider the maps

$$\Phi_n(x, y) = \big(x, y(1 + \phi_n(x, y)\big), \quad \text{with } 1 + \phi_n = \prod_{i=1}^{n} (1 + l \circ f^{i-1})^{d^{-i}} \tag{2.31}$$

and

$$l(x, y) = \frac{yk(x, y)}{1 + \widetilde{h}(x)}.$$

Then we have that

$$\begin{aligned}
\big(\Phi_n \circ f\big)_2(x, y) &= x^c y^d \big(1 + h(x, y)\big) \prod_{i=1}^{n} \big(1 + l \circ f^i(x, y)\big)^{d^{-i}} \\
&= x^c y^d \big(1 + \widetilde{h}(x)\big) \prod_{i=1}^{n+1} \big(1 + l \circ f^{i-1}(x, y)\big)^{d^{1-i}} \\
&= x^c y^d \big(1 + \widetilde{h}(x)\big) \phi_{n+1}(x, y) \\
&= \big(\widetilde{f} \circ \Phi_{n+1}\big)_2(x, y).
\end{aligned}$$

Hence if ϕ_n tends to a holomorphic map $\phi_\infty =: \phi$ in W, we are done.

But from (2.31), taking the log of the absolute value, we just have to prove that

$$\sum_{n=1}^{\infty} \left| l \circ f^{\circ n-1} \right|$$

converges in W. Since $l(x, y) = y\widetilde{l}(x, y)$, with $\widetilde{l}$ bounded in W, we can suppose, up to shrinking W, that $|l(x, y)| < y\,|y|$. Using Lemma 2.6.10, we have

$$\sum_{n=1}^{\infty} \left| l \circ f^{\circ n-1} \right| \leq 2 \sum_{n=0}^{\infty} |y_n| \leq 2\,|y_0| \sum_{n=0}^{\infty} \big(2\,|x_0|^c\big)^n,$$

that converges for $|x_0| < 1/2$ (hence for $R > 1$). □

Remark 2.6.12. With the same techniques, we can also suppose that $\widetilde{h}(x)$ in Theorem 2.6.11 does not have terms of degree $> k$, in the sense that we can erase all terms divisible by x^{k+1}, up to logarithms. Indeed, let us suppose that f is on the form of (2.30), and write $\widetilde{h}(x) =$

$\widetilde{h}_1(x) + \widetilde{h}_2(x)$, where $\widetilde{h}_1(x) = O(x^k)$ (up to logarithms) and $\widetilde{h}_2(x)$ with terms of degree $\geq k+1$ up to logarithms. Let us denote

$$\widetilde{f}(x, y) = \left(\frac{x}{\sqrt[k]{1 + x^k}}, x^c y\big(1 + \widetilde{h}_1(x)\big) \right),$$

and consider the maps

$$\Phi_n(x, y) = \big(x, y(1 + \phi_n(x, y)\big), \quad \text{with } 1 + \phi_n = \prod_{i=1}^{n} (1 + l \circ f^{i-1})^{d^{-i}}$$

and

$$l(x, y) = l(x) = \frac{\widetilde{h}_2(z)}{1 + \widetilde{h}_1(x)}.$$

As usual Φ_∞ is the conjugation we want, up to showing the convergence of ϕ_n to $\phi_\infty =: \phi$.

Taking the log of the absolute value, we just have to prove that

$$\sum_{n=1}^{\infty} \left| l \circ f^{\circ n-1} \right|$$

converges in W. Thanks to our assumptions on $\widetilde{h}_1$ and $\widetilde{h}_2$, there exists $C > 0$ such that $|l(x)| < C\,|x|^{k+1-\varepsilon}$ for a ε small enough. Using again Lemma 2.6.10, we have that $|x_n| \leq C' n^{-1/k}$ for $|x_0|$ small enough. Then we have

$$\sum_{n=1}^{\infty} \left| l \circ f^{\circ n-1} \right| \leq C \sum_{n=0}^{\infty} |x_n|^{k+1-\varepsilon} \leq C \left(C' \frac{1}{n^{1/k}} \right)^{k+1-\varepsilon},$$

that converges (for $\varepsilon < 1$).

2.7. Normal forms

2.7.1. Nilpotent case

Favre and Jonsson studied the superattracting case (see [27, Theorem 5.1]); the nilpotent case is almost the same, there is in fact just one little difference between them (see Remark 2.7.4).

Lemma 2.7.1. *Let f be a (dominant) holomorphic germ, with df_0 non-invertible, $\nu_\star$ an eigenvaluation for f, and $(\pi, p, \hat{f})$ a rigidification obtained from $\nu_\star$ as in Theorem 2.3.2.*

Assume $\nu_\star$ is not a divisorial valuation. Then $c_\infty(\hat{f}) = c_\infty(f)$.

Proof. Directly from the definition of $\hat{f}$ as lift of f, we have $\pi \circ \hat{f} = f \circ \pi$. Let $\mu_\star = \pi_\bullet^{-1}(\nu_\star)$ (in this case $\mu_\star$ is an eigenvaluation for $\hat{f}$). Then:

$$\begin{array}{l} c(\pi \circ \hat{f}, \mu_\star) = c(\hat{f}, \mu_\star) \cdot c(\pi, \hat{f}_\bullet \mu_\star) = c(\hat{f}, \mu_\star) \cdot c(\pi, \mu_\star) \\ \quad \| \\ c(f \circ \pi, \mu_\star) = c(\pi, \mu_\star) \cdot c(f, \pi_\bullet \mu_\star) = c(\pi, \mu_\star) \cdot c(f, \nu_\star). \end{array}$$

From Theorem 2.2.16 we have $c(f, \nu_\star) = c_\infty(f)$ and $c(\hat{f}, \mu_\star) = c_\infty(\hat{f})$; so, if $c(\pi, \mu_\star) < \infty$, we have $c_\infty(f) = c_\infty(\hat{f})$. But $c(\pi, \mu_\star) = \infty$ if and only if $\mu_\star \in \partial U(p)$; following the proof of Theorem 2.3.2, this (always) happens if and only if $\nu_\star$ is a divisorial valuation. □

Lemma 2.7.2. *Let $f : (\mathbb{C}^2, 0) \to (\mathbb{C}^2, 0)$ be a holomorphic germ. Then df_0 is nilpotent if and only if $c_\infty(f) > 1$.*

Proof. If df_0 is nilpotent, then $df_0^2 = 0$, and $c(f^2) \geq 2$; we have $c(f^{2n}) \geq 2^n$, and $c_\infty(f) \geq \sqrt{2}$. On the contrary, if df_0 is not nilpotent, then $df_0^n \neq 0$ for every $n \in \mathbb{N}$, and then $c(f^n) = 1$, that implies also $c_\infty(f) = 1$. □

Theorem 2.7.3. *Let f be a (dominant) nilpotent holomorphic germ, $\nu_\star$ an eigenvaluation for f, and $(\pi, p, \hat{f})$ a rigidification obtained from $\nu_\star$ as in Theorem 2.3.2. Then:*

(i) *if $\nu_\star$ is a (non-contracted) analytic curve valuation, then $\hat{f} \cong (z^a, z^c w^d)$, with $d \geq a \geq 2$, $c \geq 1$, $a = c_\infty(f)$;*
(ii) *$\nu_\star$ cannot be a non-analytic curve valuation;*
(iii) *if $\nu_\star$ is an infinitely singular valuation, then $\hat{f} \cong (z^a, \alpha z^c w + P(z))$, with $2 \leq a \in \mathbb{N}$, $c \geq 1$, $\alpha \neq 0$, and $P \in z\mathbb{C}[z]$ ($P \not\equiv 0$);*
(iv) *if $\nu_\star$ is a divisorial valuation, then $\hat{f} \cong (z^a w^b, \alpha w)$, with $a = c_\infty(f) \geq 2$, $b \geq 1$ and $\alpha \neq 0$;*
(v) *if $\nu_\star$ is an irrational valuation, then either $\hat{f} \cong (z^n, w^n)$ with $2 \leq n = c_\infty(f) \in \mathbb{N}$, or $\hat{f} \cong (w^b, z^c)$ with $b, c \geq 1$ and $c_\infty(f) = \sqrt{bc} \notin \mathbb{Q}$, or $\hat{f} \cong (\lambda_1 z^a w^b, \lambda_2 z^c w^d)$, with $a, b, c, d \geq 1$, $ad - bc \neq 0$, $\lambda_1, \lambda_2 \neq 0$ (we can suppose $\lambda_1 = \lambda_2 = 1$ if $bc \neq (a-1)(d-1)$); $c_\infty(f)$ is an eigenvalue of the matrix with entries (a, b, c, d), and it is the larger one if and only if $c(\hat{f})$ is greater than the smaller one.*

Proof. We split as usual the proof into five cases, depending on the type of $\nu_\star$.

(i) If $\nu_\star = \nu_C$ is a (non-contracted) analytic curve valuation, then we have seen in part (i) of the proof of Theorem 2.3.2 that $\mathcal{C}^\infty(\hat{f}) = E \cup \tilde{C}$ or $\mathcal{C}^\infty(\hat{f}) = E$, and in both cases E is contracted and $\tilde{C}$ is fixed by $\hat{f}$. First of all, we know from Lemma 2.7.1 and Lemma 2.7.2 that $\hat{f}$ is nilpotent, and so $\operatorname{tr} d\hat{f}_p = 0$.
In the first case $\mathcal{C}^\infty(\hat{f}) = E \cup \tilde{C}$ is reducible: so $\hat{f}$ is of class 6 or 7. We can choose (z, w) local coordinates in p such that $E = \{z = 0\}$, $\tilde{C} = \{w = 0\}$, and

$$\hat{f}(z, w) = \Big(\lambda_1 z^a w^b \big(1 + \phi_1(z, w)\big), \lambda_2 z^c w^d \big(1 + \phi_2(z, w)\big)\Big),$$

with $\phi_1(0, 0) = \phi_2(0, 0) = 0$, and $\lambda_1, \lambda_2 \neq 0$.
Let us denote by $M = M(\hat{f})$ the 2×2 matrix as in (2.5): $\hat{f}$ is of class 6 if and only if $\det M \neq 0$.
Since E is contracted, then $a, c \geq 2$, while $\tilde{C}$ fixed implies $b = 0$ and $d \geq 1$. So M is invertible (triangular) and $\hat{f}$ is of class 6. By Theorem 2.4.3, up to local conjugation we can suppose $\phi_1 = \phi_2 = 0$. Thanks to Lemma 2.7.1, Theorem 2.2.16 and Proposition 2.2.6, we know that $c_\infty(f) = c_\infty(\hat{f}) = c(\hat{f}, \nu_{\tilde{C}}) = e(\hat{f}, \tilde{C}) = a$, and then we have $a \geq 2$, and $d \geq 2$ (being $a = c_\infty(\hat{f}) = \min\{a, d\}$). In particular $1 \notin \operatorname{Spec}(M)$, and we can also suppose $\lambda_1 = \lambda_2 = 1$.
Putting together all the informations, we have

$$\hat{f}(z, w) \cong (z^a, z^c w^d),$$

with $d \geq a \geq 2, c \geq 1$ (and $a = c_\infty(f)$).
In the second case, $\mathcal{C}^\infty(\hat{f}) = E$ is irreducible, so $\hat{f}$ is of class 4. By Theorem 2.4.3, we can choose local coordinates (z, w) such that $E = \{z = 0\}, \tilde{C} = \{w = 0\}$, and

$$\hat{f}(z, w) = \big(z^a, \alpha z^c w + P(z)\big),$$

with $a \geq 2$ and $c \geq 1$. Since $\tilde{C}$ is fixed, then $P \equiv 0$. But in this case, there is a monomial eigenvaluation ν, defined by $\nu(z) = a - 1$ and $\nu(w) = c$ (and opportunely renormalized): a contradiction, since $f_\bullet^n \to \nu_w$ on $\mathcal{V} \setminus \{\nu_z\}$ (see Remark 2.3.3 and Theorem 2.3.2).

(ii) Corollary 2.2.20.

(iii) If $\nu_\star$ is an infinitely singular valuation, then we have seen in part (iii) of the proof of Theorem 2.3.2 that $\mathcal{C}^\infty(\hat{f}) = E$ and $\hat{f}(E) = p$. Again, we know from Lemma 2.7.1 and Lemma 2.7.2 that $\hat{f}$ is nilpotent, and so $\operatorname{tr} d\hat{f}_p = 0$; moreover $\mathcal{C}^\infty(\hat{f})$ is irreducible, and then $\hat{f}$ is of class 4. So we can choose (z, w) local coordinates in p such that $E = \{z = 0\}$, and

$$\hat{f}(z, w) = \big(z^a, \alpha z^c w + P(z)\big),$$

with $a \geq 2$ and $c \geq 1$. Since $f_\bullet^n \to \nu_\star$ in $U(p)$, no curve valuation is fixed by $\hat{f}$, and $P \not\equiv 0$. Moreover, $c_\infty(f) = a$ (see [27, p. 25]).

(iv) If $\nu_\star = \nu_E$ is a divisorial valuation, then we have seen in part (iv) of the proof of Theorem 2.3.2 that $\mathcal{C}^\infty(\hat{f}) = E \cup F$, with $E, F \in \Gamma_\pi^*$ two adjacent exceptional components of a modification π. Moreover, F is contracted to p by $\hat{f}$, while E is fixed. In this case we cannot apply Lemma 2.7.1, but we know that $\mathcal{C}^\infty(\hat{f})$ is reducible, and so we can choose (z, w) local coordinates in p such that $E = \{z = 0\}$, $F = \{w = 0\}$, and

$$\hat{f}(z, w) = \Big(\lambda_1 z^a w^b\big(1 + \phi_1(z, w)\big), \lambda_2 z^c w^d\big(1 + \phi_2(z, w)\big)\Big),$$

with $\phi_1(0, 0) = \phi_2(0, 0) = 0$, and $\lambda_1, \lambda_2 \neq 0$. Since E is fixed by $\hat{f}$, then $c = 0$ (and $a \geq 1$); since F is contracted, we have $b, d \geq 1$. Choosing a non-critical fixed point of $df_\bullet$ (see Remark 2.2.15) in Proposition 2.2.21, we can also suppose $d = 1$. Using notations as in point (i), we have $\mathrm{tr}(d\hat{f}_0) = \lambda_2 \neq 0$, so $\hat{f}$ is of class 5, and

$$\hat{f} \cong (z^a w^b, \alpha w),$$

with $\alpha = \lambda_2 \neq 0$, $a \geq 2$ and $b \geq 1$ (and in this case $c_\infty(\hat{f}) = 1$). From direct computations (using Lemma 2.2.19) it follows that $a = c_\infty(f)$.

(v) Let us finally suppose $\nu = \nu_\star$ an irrational valuation. As we have seen in part (v) of the proof of Theorem 2.3.2, $\mathcal{C}^\infty(\hat{f}) = E \cup F$, with $E, F \in \Gamma_\pi^*$ two adjacent exceptional components of the modification π. Recalling Lemma 2.7.1, we know that $\hat{f}$ is of class 6 or 7, and we can choose (z, w) local coordinates in p such that $E = \{z = 0\}$, $F = \{w = 0\}$, and

$$\hat{f}(z, w) = \Big(\lambda_1 z^a w^b\big(1 + \phi_1(z, w)\big), \lambda_2 z^c w^d\big(1 + \phi_2(z, w)\big)\Big),$$

with $\phi_1(0, 0) = \phi_2(0, 0) = 0$, and $\lambda_1, \lambda_2 \neq 0$.
As always we denote by $M = M(\hat{f})$ the 2×2 matrix of entries (a, b, c, d) as in (2.5).
For the dynamics on $\mathcal{C}^\infty(\hat{f})$, we have seen in Proposition 2.2.21 that there are three cases.

Let us denote $I = [\nu_E, \nu_F]$, $U = U(p)$, and $J = [\nu_z, \nu_w]$: by our choices we have $\pi_\bullet(J) = I$. If $f_\bullet|_I = \mathrm{id}_I$, then both E and F are fixed by $f_\bullet$; in this case $b = c = 0$, while $a, d \geq 2$. Then $\det M \neq 0$ (M is diagonal), and $1 \notin \mathrm{Spec}(M)$, so $\hat{f}$ is of class 6, and conjugated to (z^a, w^d). Moreover, $f_\bullet|_I = \mathrm{id}_I$ implies $\hat{f}_\bullet = \mathrm{id}_J$, and this easily implies $a = d(= c_\infty(f))$. So in this case we have

$$\hat{f} \cong (z^n, w^n),$$

with $2 \leq n = c_\infty(f) \in \mathbb{N}$.
If $f_\bullet|_I{}^2 = \mathrm{id}_I$ (but $f_\bullet|_I \neq \mathrm{id}_I$), then E and F are exchanged by $f_\bullet$; in this case $a = d = 0$, while $b, c \geq 1$ (not both $= 1$). Then $\det M \neq 0$, and $1 \notin \mathrm{Spec}(M)$, so $\hat{f}$ is of class 6, and

$$\hat{f} \cong (w^b, z^c),$$

with $c_\infty(f) = \sqrt{bc}$.
Moreover, we know that $\mu_\star := \pi_\bullet^{-1}\nu_\star$ is an irrational valuation (see Proposition 2.2.4), and from direct computations we get $\sqrt{bc} = c_\infty(f) \notin \mathbb{Q}$.
Finally if $f_\bullet|_I{}^2 \neq \mathrm{id}_I$, then $f_\bullet U \Subset U$, and both E and F are contracted to p. This time we have $a, b, c, d \geq 1$; we claim that $\det M \neq 0$ in this case too. For every $s, t \geq 0$, let us consider the monomial valuations $\mu_{s,t}$ in the local coordinates (z, w) defined by $\mu_{s,t}(z) = s$ and $\mu_{s,t}(w) = t$. Then (see Remark 2.3.4), I is parametrized by $\nu_{s,t} = \pi_*\mu_{s,t}$, with s, t satisfying $b_E s + b_F t = 1$. It is easy to see that $b_E\nu_E = \nu_{0,1}$ and $b_F\nu_F = \nu_{1,0}$, while $\hat{f}_*\mu_{1,0} = \mu_{a,b}$, and $\hat{f}_*\mu_{0,1} = \mu_{c,d}$. Since $f_\bullet$ is injective in I, we have that $\mu_{a,b}$ and $\mu_{c,d}$ are not proportional, that is to say $ad \neq bc$ and $\det M \neq 0$. So we have again $\hat{f}$ of class 6, and:

$$\hat{f} \cong (\lambda_1 z^a w^b, \lambda_2 z^c w^d),$$

where $a, b, c, d \geq 1$, $ad \neq bc$ and we can suppose $\lambda_1 = \lambda_2 = 1$ if $(a-1)(d-1) \neq bc$. A direct computation shows that $c_\infty(f) \in \mathrm{Spec}(M)$, and it is the larger eigenvalue if and only if $c(\hat{f})$ is greater than the smaller eigenvalue. □

Remark 2.7.4. The unique difference between the superattracting case and the nilpotent case is that, in the nilpotent case, one has $c_\infty(f) \geq \sqrt{2}$, while in the superattracting case one has $c_\infty(f) \geq 2$. Moreover, thanks to Lemma 2.7.1, when the eigenvaluation $\nu_\star$ is not divisorial, then for the lift $\hat{f}$ we have $c_\infty(f) = c_\infty(\hat{f})$. So to obtain the result for the nilpotent case, we have just to ignore the hypothesis $c_\infty(\hat{f}) \geq 2$ (when $\nu_\star$ is not divisorial).

2.7.2. Semi-superattracting case

Germs of type $(0, \mathbb{D}^*)$

Proposition 2.7.5. *Let f be a (dominant) holomorphic germ of type $(0, \mathbb{D}^*)$, $\nu_\star$ an eigenvaluation for f, and $(\pi, p, \hat{f})$ a rigidification obtained from $\nu_\star$ as in Theorem 2.3.2. Let $\lambda \in \mathbb{D}^*$ be the non-zero eigenvalue of df_0. Then $\nu_\star$ can be only a (formal) curve valuation, and:*

(i) *if $\nu_\star$ is a (non-contracted) analytic curve valuation, then $\hat{f} \cong (\lambda x, x^c y^d)$, with $c \geq 1$ and $d \geq 1$;*

(ii) *if $\nu_\star$ is a non-analytic curve valuation, then $\hat{f} \cong (\lambda x, x^q y + P(x))$, with $q \geq 1$, and $P \in x\mathbb{C}[x]$ with* $\deg P \leq q$, $P \not\equiv 0$.

Proof. The first assertion follows from Theorem 2.3.8.

(i) If $\nu_\star = \nu_C$ is a (non-contracted) analytic curve valuation, then directly from Theorem 2.7.3 we have that $\mathcal{C}^\infty(\hat{f}) = E \cup \tilde{C}$ or $\mathcal{C}^\infty(\hat{f}) = E$, and in both cases E is contracted and $\tilde{C}$ is fixed by $\hat{f}$. We also know from Proposition 2.3.6 that $\operatorname{tr} d\hat{f}_p = \lambda \neq 0$.
In the first case $\mathcal{C}^\infty(\hat{f}) = E \cup \tilde{C}$ is reducible: so $\hat{f}$ is of class 5. Hence we can choose local coordinates (x, y) in p such that $E = \{x = 0\}$, $\tilde{C} = \{y = 0\}$, and
$$\hat{f}(x, y) = (\lambda x, x^c y^d),$$
with $c \geq 1$ and $d \geq 2$.
In the second case $\mathcal{C}^\infty(f) = E$ is irreducible: so $\hat{f}$ is of class 2 or 3, but since E is contracted to 0 by $\hat{f}$, then $\hat{f}$ is of class 2. Hence we can choose local coordinates (x, y) such that $E = \{x = 0\}$, $\tilde{C} = \{y = 0\}$, and
$$\hat{f}(x, y) = \big(\lambda x, x^q y + P(x)\big),$$
with $q \geq 1$. Since $\tilde{C}$ is fixed, then $P \equiv 0$.

(ii) Let us suppose $\nu_\star = \nu_C$ a non-analytic curve valuation. We have seen in the proof of Theorem 2.3.2 that $\mathcal{C}^\infty(\hat{f}) = E$, and E is contracted to 0 by $\hat{f}$. We also know from Proposition 2.3.6 that $\operatorname{tr} d\hat{f}_p = \lambda \neq 0$, so $\hat{f}$ is of class 2 or 3. But only for maps in class 2 $\hat{f}$ contracts the component E in $\mathcal{C}^\infty(\hat{f})$. So we are in class 2 and we can choose local coordinates (x, y) at p such that $E = \{x = 0\}$, and
$$\hat{f}(x, y) = \big(\lambda x, x^q y + P(x)\big),$$
with $q \geq 1$, and $P \in x\mathbb{C}[x]$ with $\deg P \leq q$. Since $f_\bullet^n \to \nu_\star$ in $U(p)$, no analytic curve valuation (besides ν_x) is fixed by $\hat{f}$, and $P \not\equiv 0$. □

Germs of type $(0, \mathbb{C} \setminus \overline{\mathbb{D}})$

Proposition 2.7.6. *Let f be a (dominant) holomorphic germ of type $(0, \mathbb{C} \setminus \overline{\mathbb{D}})$, $\nu_\star$ an eigenvaluation for f, and $(\pi, p, \hat{f})$ a rigidification obtained from $\nu_\star$ as in Theorem 2.3.2. Let $\lambda \in \mathbb{C} \setminus \overline{\mathbb{D}}$ be the non-zero eigenvalue of df_0. Then $\nu_\star$ can be only an analytic curve valuation, and $\hat{f} \overset{\text{for}}{\cong} (\lambda x, x^c y^d(1 + \varepsilon x^l))$, with $c \geq 1, d \geq 1, l \geq 1$ and $\varepsilon = 0$ if $\lambda^l \neq d$, or $\varepsilon \in \{0, 1\}$ if $\lambda^l = d$.*

Proof. Thanks to Theorem 2.3.8, we know that $\nu_\star$ has to be a (formal) curve valuation.

Let us suppose $\nu_\star = \nu_C$ a non-analytic curve valuation. From the proof of Theorem 2.3.2 and Proposition 2.2.21.(i) we know that $f_\bullet^n \to \nu_C$ on a suitable open set $U = U(p)$, and hence $\hat{f}_\bullet^n \to \nu_{\tilde{C}}$ on $\mathcal{V} \setminus \nu_E$, where $\tilde{C}$ is the strict transform of C (and it is non-analytic as well). Notice that ν_E is an analytic curve valuation if considered on the valuative tree where $\hat{f}_\bullet$ acts. In particular, E is the only analytic curve fixed by $\hat{f}$, that is in contradiction with Theorem 1.1.33 (Stable-Unstable Manifold Theorem), since we know from Proposition 2.3.6 that $\mathrm{Spec}(d\hat{f}_p) = \{0, \lambda\}$ and $|\lambda| > 1$. So $\nu_\star = \nu_C$ is a (non-contracted) analytic curve valuation.

Then the assertion on normal forms follows from Theorem 2.4.7. □

Germs of type $(0, \partial\mathbb{D})$

Proposition 2.7.7. *Let f be a (dominant) holomorphic germ of type $(0, \partial\mathbb{D})$, $\nu_\star$ an eigenvaluation for f, and $(\pi, p, \hat{f})$ a rigidification obtained from $\nu_\star$ as in Theorem 2.3.2. Let $\lambda \in \partial\mathbb{D}$ be the non-zero eigenvalue of df_0. Then $\nu_\star$ can be only a (formal) curve valuation, and:*

(i) *if λ is not a root of unity, then $\hat{f} \overset{\text{for}}{\cong} (\lambda x, x^c y^d)$, with $c, d \geq 1$;*
(ii) *if $\lambda^r = 1$ is a root of unity, then $\hat{f} \overset{\text{for}}{\cong} \big(\lambda x(1 + x^s + \beta x^{2s}), x^c y^d (1 + \varepsilon(x^r))\big)$, where $c, d \geq 1, r|s$ and $\beta \in \mathbb{C}$, while ε is a formal power series in x^r, and $\varepsilon \equiv 0$ if $d \geq 2$*

Proof. The first assertion follows from Theorem 2.3.8, while the normal forms are given by Theorem 2.4.7. □

2.7.3. Some remarks and examples

Remark 2.7.8. The proof of Theorem 2.3.2 gives a general procedure to obtain a rigid germ. But in specific instances we can choose an infinitely near point lower that the one indicated. In particular, if $\nu_\star$ is divisorial, it can happen that $U = U(p)$ can be associated to a free point p, and not to a satellite one. If this is the case, we obtain a irreducible rigid germ, of class 2 or 3; and it has to be of class 3, since the generalized critical set E is invariant but not contracted by $\hat{f}$. So, for example, if $\hat{f}$ is still attracting, then $\hat{f} \cong (z^p, \alpha w)$, with $p \geq 2$ and $0 < |\alpha| < 1$ (with $\alpha = \lambda$ if $df_0 = D_\lambda$).

Example 2.7.9. We present an example of the phenomenon we described in Remark 2.7.8. Set

$$f(z, w) = (z^n + w^n, w^n),$$

with $n \geq 2$ an integer. We easily see that $\nu_{\mathfrak{m}}$ is an eigenvaluation for f. We want to study the action of $\hat{f}$ on the exceptional component $E = E_0$ that arises from the single blow-up of the origin. We can study it by checking the action of $f_\bullet$ on $\mathcal{E} := \{\nu_{y-\theta x} \mid \theta \in \mathbb{C}\} \cup \{\nu_x\}$, where we fix the correspondence $\theta \mapsto \nu_{y-\theta x}$ between $E \cong \mathbb{P}^1(\mathbb{C})$ and $\mathcal{E}$ (setting $\infty \mapsto \nu_x$). Direct computations show that

$$\hat{f}|_E : \theta \mapsto \frac{\theta^n}{1+\theta^n}.$$

Now set $p = \theta \in E$ such that θ is a non-critical fixed point for $\hat{f}|_E$, *i.e.*, such that $\theta^n + 1 = \theta^{n-1}$, and lift f to a holomorphic germ $\hat{f}$ on the infinitely near point p. Using the same arguments as in the proof of Theorem 2.3.2, we can tell that $\hat{f}$ is a rigid germ.

We show this claim by direct computations. Let us make a blow-up in $0 \in \mathbb{C}^2$:

$$\begin{cases} z = u, \\ w = ut; \end{cases} \qquad \begin{cases} u = z, \\ t = w/z; \end{cases}$$

we obtain

$$\hat{f}(u, t) = \left(u^n(1 + t^n), \frac{t^n}{1+t^n}\right).$$

Choosing the local coordinates $(u, v := t - \theta)$, we obtain

$$\hat{f}(u, v) = \Big(u^n\big(1 + (v + \theta)^n\big), v\xi(v)\Big),$$

for a suitable invertible germ ξ. In particular, $\hat{f}$ is a rigid germ, it belongs to class 3, and (by direct computation) it is locally holomorphically conjugated to $(u, v) \mapsto (u^n, \alpha v)$, for a suitable $\alpha \neq 0$, whereas Theorem 2.7.3 would give us a germ that belongs to class 5. In this case, we recover the result of Theorem 2.7.3 simply by taking the lift of $g = \hat{f}$ when we blow-up the point $[0 : 1] \in E$, and obtaining

$$\hat{g}(x, y) = \big(x^n y^{n-1}\chi(y), v\xi(v)\big),$$

for a suitable invertible germ χ; this germ is locally holomorphically conjugated to $(x^n y^{n-1}, \alpha y)$.

Remark 2.7.10. We can apply Theorem 2.7.3, Propositions 2.7.5, 2.7.6 and 2.7.7 even when f is rigid itself: the result is that we can avoid some kind of rigid germs. First of all, from the proof of Theorem 2.7.3 (and recalling Proposition 2.3.6), one can see that Class 7 can be always avoided (hence Class 7 is not "stable under blow-ups"). Moreover, from the proof of Theorem 2.3.2, we see that the germs we obtain after lifting are such that $\hat{f}_\bullet$ has always only one fixed point $\mu_\star = \pi_\bullet^{-1}\nu_\star$ of the same type of $\nu_\star$, with two exceptions: either $\nu_\star$ is divisorial, and $\mu_\star$ turns out to be an analytic curve valuation (contracted by π), or $\nu_\star$ is an irrational eigenvaluation, and in this case it can happen that $\hat{f}_\bullet = \mathrm{id}$ on $[\nu_z, \nu_w]$.

In the first case reapplying Propositions 2.7.5, 2.7.6 and 2.7.7 we see that we obtain the same type of germ. In the second case, we have, up to local holomorphic conjugacy, that $\hat{f}(z, w) = (z^n, w^n)$, with a suitable $n \geq 2$. Then all valuations on $[\nu_z, \nu_w]$ are eigenvaluations, and reapplying Theorem 2.7.3, we obtain a rigid germ that belongs to a different class. In particular, making a single blow-up on the origin, and considering the germ at $[1 : 1]$, we obtain a germ of the form $(nz(1 + h(z)), w^n)$ for a suitable holomorphic map h such that $h(0) = 0$, that is (by direct computation) holomorphically conjugated to (nz, w^n).

Example 2.7.11. Reapplying Theorem 2.7.3, as just seen in the last remark, we usually obtain the same normal form type. But there are cases where the normal form can change (staying rigid).

Consider for example the rigid germ $f(z, w) = (w^2, z^3)$. Then the only eigenvaluation $\nu_\star$ is the monomial valuation on (z, w), such that $\nu_\star(x) = 1$ and $\nu_\star(w) = \sqrt{3/2}$. Then an infinitely near point p that works in Theorem 2.7.3 can be obtained after three blow-ups: the first at 0 (and we obtain E_0), the second at $[1 : 0] \in E_0$ (and we obtain E_1), the third at $[0 : 1]$ (and we obtain E_2). We can choose $p = [0, 1] \in E_2$, and the lift we obtain is $\hat{f}(z, w) = (w^6, z)$.

Chapter 3
Rigid germs in higher dimension

3.1. Definitions

3.1.1. Notations

In this section, we shall specify all the notations we are going to use for formal power series in this chapter.

First of all, for variables or functions in all formulas, we shall denote by bold letters the vectors, such as $\mathbf{x}$ or $\mathbf{f}$, and by standard letters the coordinates; for example $\mathbf{x} = (x_1, \dots, x_d)^T$.

In this chapter, we shall be careful to distinguish between horizontal and vertical vectors. So vectors in $\mathbb{C}^d$ shall be vertical vectors, while differentials of maps $\mathbf{f} : (\mathbb{C}^k, 0) \to (\mathbb{C}^h, 0)$ shall be a matrix $d\mathbf{f}_0 \in \mathcal{M}(h \times k, \mathbb{C})$.

Definition 3.1.1. Let $\mathbf{x} = (x_1, \dots, x_k)^T$ and $\underline{A} = (a_i^j) \in \mathcal{M}(h \times k, \mathbb{Q})$. Then we shall denote

$$\mathbf{x}^{\underline{A}} = \big((\mathbf{x}^{\underline{A}})_1, \dots (\mathbf{x}^{\underline{A}})_h\big)^T, \qquad (\mathbf{x}^{\underline{A}})_i = \prod_{j=1}^{k} x_j^{a_i^j}.$$

Notice that if $\mathbf{a}_i := \mathbf{e}_i \underline{A}$ is the i-th row of $\underline{A}$, where $\mathbf{e}_i$ is the i-th element of the canonical (dual) basis, then $(\mathbf{x}^{\underline{A}})_i = \mathbf{x}^{\mathbf{a}_i}$.

Definition 3.1.2. Let $f(\mathbf{x})$ be a formal power series in d (complex) variables $\mathbf{x} = (x_1, \dots, x_d)^T$. Than we shall write

$$f(\mathbf{x}) = \sum_{\mathbf{i} \in \mathbb{N}^d} f_{\mathbf{i}} \mathbf{x}^{\mathbf{i}},$$

where $\mathbf{i}$ are (horizontal) vectors $\mathbf{i} = (i^1, \dots, i^d)$.

If we have $\mathbf{f}(\mathbf{x}) = \big(f_1(\mathbf{x}), \dots, f_c(\mathbf{x})\big)^T$ we shall denote

$$f_k(\mathbf{x}) = \sum_{\mathbf{i} \in \mathbb{N}^d} f_{k,\mathbf{i}} \mathbf{x}^{\mathbf{i}}.$$

Definition 3.1.3. We shall denote by $\mathfrak{m}$ the maximal ideal of $\mathbb{C}[[\mathbf{x}]]$, with $\mathbf{x} = (x_1, \ldots, x_d)^T$. In particular $\mathfrak{m}$ is generated by $x_1, \ldots, x_d$, and it is the set of formal power series f such that $f(\mathbf{0}) = f_\mathbf{0} = 0$. If we would like to specify the number of coordinates, we shall write $\mathfrak{m}_d$.

Definition 3.1.4. In the following, we shall use a few non-standard notations, that are going to be specified here. If $\lambda \in \mathbb{C}$ and $\underline{A} \in \mathcal{M}(k \times h, \mathbb{C})$ is a matrix, then as usual we shall denote by

$$\lambda \underline{A} \in \mathcal{M}(k \times h, \mathbb{C})$$

the product component by component between λ and every element of $\underline{A}$.

Let $\mathbf{a} \in \mathcal{M}(k \times 1, \mathbb{C})$ be a vertical vector, and $\mathbf{b} \in \mathcal{M}(k \times h, \mathbb{C})$ be a matrix. We shall denote by

$$\mathbf{a} : \mathbf{b} = \mathbf{b} : \mathbf{a} \in \mathcal{M}(k \times h, \mathbb{C}), \qquad (\mathbf{a} : \mathbf{b})_i^j = a_i b_i^j$$

the product component by component between elements of $\mathbf{a}$ and rows of $\mathbf{b}$. When $h = 1$, we shall omit the product symbol $:$ and simply write $\mathbf{ab}$.

Analogously, let $\mathbf{a} \in \mathcal{M}(1 \times k, \mathbb{C})$ be a horizontal vector, and $\mathbf{b} \in \mathcal{M}(h \times k, \mathbb{C})$ be a matrix. We shall denote by

$$\mathbf{a} \cdot\cdot\, \mathbf{b} = \mathbf{b} \cdot\cdot\, \mathbf{a} \in \mathcal{M}(h \times k, \mathbb{C}), \qquad (\mathbf{a} \cdot\cdot\, \mathbf{b})_i^j = a^j b_i^j$$

the product component by component between elements of $\mathbf{a}$ and columns of $\mathbf{b}$.

Finally, let $\underline{A} \in \mathcal{M}(h \times k, \mathbb{C})$ and $\underline{B} \in \mathcal{M}(k \times l, \mathbb{C})$ be two matrices, then we shall denote by

$$\underline{A} \cdot \underline{B} \in \mathcal{M}(h \times l, \mathbb{C})$$

the standard matrix product. When $h = 1$ or $l = 1$, and hence we are considering the standard product between a matrix and a vector, we shall omit the product symbol $\cdot$.

We notice that depending on the dimension of factors, when we omit the product symbol, only one between $\cdot$ and $:$ is defined, unless we are in some trivial cases where the two products coincide (namely, when $k = 1$ in both cases).

For sums, if $\underline{A}, \underline{B} \in \mathcal{M}(k \times h, \mathbb{C})$, we shall as usual denote by

$$\underline{A} + \underline{B} \in \mathcal{M}(k \times h, \mathbb{C})$$

the standard sum component by component.

Let $\lambda \in \mathbb{C}$ be a complex number and $\underline{A} \in \mathcal{M}(k \times h, \mathbb{C})$ be a matrix. Then we shall denote by

$$\lambda + \underline{A} \in \mathcal{M}(k \times h, \mathbb{C})$$

the matrix obtained from $\underline{A}$ by adding λ to each entry.

Remark 3.1.5. Let us consider the map $f : \mathbb{C}^k \to \mathbb{C}^h$ given by

$$\mathbf{f}(\mathbf{x}) = \mathbf{x}^{\underline{A}},$$

where $\underline{A} \in \mathcal{M}(h \times k, \mathbb{N})$. We want to compute $d\mathbf{f}_{\mathbf{x}} \in \mathcal{M}(h \times k, \mathbb{C})$ for every $\mathbf{x} \in \mathbb{C}^k$.

First of all $\mathbf{f} = (f_1, \dots, f_h)^T$, where

$$f_i(\mathbf{x}) = \mathbf{x}^{\mathbf{a}_i} = \prod_{j=1}^{k} x_j^{a_i^j}.$$

Then we have that

$$(d\mathbf{f})_i^j := \frac{\partial f_i}{\partial x_j} = a_i^j \mathbf{x}^{\mathbf{a}_i} x_j^{-1}.$$

Using our notations, we get

$$(d\mathbf{f})_i = \mathbf{x}^{\mathbf{a}_i} a_i \cdot\cdot \left(\mathbf{x}^{-\underline{I}}\right)^T,$$

where we are considering the product between a scalar, a vector $1 \times k$, and another vector $1 \times k$, coordinate by coordinate.

For the matrix expression, we get then

$$d\mathbf{f} = \underline{A} : \mathbf{x}^{\underline{A}} \cdot\cdot \left(\mathbf{x}^{-\underline{I}}\right)^T,$$

where we are considering this time the product between a matrix $h \times k$, a vector $h \times 1$, and a vector $1 \times k$. We notice that in this case we have to apply the products from left to right to make them have sense, but we would obtain the same result by multiplying the first and the second factor as matrices, and then multiplying the two matrices $h \times k$ component by component.

Definition 3.1.6. Let $f(\mathbf{x})$ be a formal power series in d (complex) variables $\mathbf{x} = (x_1, \dots, x_d)^T$. Let $I = (\mathbf{i}_1, \dots, \mathbf{i}_n)^T \in \mathcal{M}(n \times d, \mathbb{N})$ be a vector of multi-indices. Then we shall denote by f_I the product

$$f_I = \prod_{l=1}^{n} f_{\mathbf{i}_l}.$$

We shall denote by $|I|$ the vector in $\mathcal{M}(1 \times d, \mathbb{N})$ given by

$$|I| = \sum_{l=1}^{n} \mathbf{i}_l.$$

Definition 3.1.7. Let $\mathbf{f}(\mathbf{x}) = (f_1, \ldots, f_r)^T$ be a r-uple of formal power series in d (complex) variables $\mathbf{x} = (x_1, \ldots, x_d)^T$. Let $\mathbf{n} = (n^1, \ldots, n^r) \in \mathcal{M}(1 \times r, \mathbb{N})$ be a multi-index, and $\mathbf{I} = (I^1, \ldots, I^r)$ be a (horizontal) vector of matrices, where $I^l \in \mathcal{M}(n^l \times d, \mathbb{N})$. Then we shall denote by $f_{\mathbf{I}}$ the product

$$f_{\mathbf{I}} = \prod_{l=1}^{r} f_{l,I^l}.$$

We shall denote by $|\mathbf{I}|$ the vector in $\mathcal{M}(1 \times d, \mathbb{N})$ given by

$$|\mathbf{I}| = \sum_{l=1}^{r} \left|I^l\right|.$$

Remark 3.1.8. Let $f(\mathbf{x})$ be a formal power series in d (complex) variables $\mathbf{x} = (x_1, \ldots, x_d)^T$, and $n \in \mathbb{N}^*$ Then

$$\big(f(\mathbf{x})\big)^n = \sum_{I} f_I \mathbf{x}^{|I|},$$

where the dummy variable I varies in $\mathcal{M}(n \times d, \mathbb{N})$.

Then suppose that we have $\mathbf{f}(\mathbf{x}) = (f_1(\mathbf{x}), \ldots, f_r(\mathbf{x}))^T$ an r-uple of formal power series in d (complex) variables $\mathbf{x} = (x_1, \ldots, x_d)^T$, and $\mathbf{n} = (n^1, \ldots n^r) \in \mathcal{M}(1 \times r, \mathbb{N})$, with $\mathbf{n} \neq \mathbf{0}$. In this case we have

$$\big(\mathbf{f}(\mathbf{x})\big)^{\mathbf{n}} = \prod_{j=1}^{r} \big(f_j(\mathbf{x})\big)^{n_j} = \sum_{\mathbf{I}} f_{\mathbf{I}} \mathbf{x}^{|\mathbf{I}|},$$

where the dummy variable is $\mathbf{I} = (I^1, \ldots, I^r)$, where $I^l \in \mathcal{M}(n^l \times d, \mathbb{N})$ for $l = 1, \ldots, r$.

If $\underline{N} \in \mathcal{M}(s \times r, \mathbb{N})$, we shall have

$$\Big(\big(\mathbf{f}(\mathbf{x})\big)^{\underline{N}}\Big)_h = \big(\mathbf{f}(\mathbf{x})\big)^{\mathbf{n}_h}$$

for $h = 1, \ldots, s$, where $\mathbf{n}_h = \mathbf{e}_h \underline{N}$ is the h-th row of $\underline{N}$. .

For the estimates in the next sections, we shall need the following Lemma and Proposition.

Lemma 3.1.9. *Let $\mathbf{f} = (f_1, \ldots, f_r)^T$ be an r-uple (of formal power series), and let $\underline{D} \in \mathcal{M}(s \times r, \mathbb{Q})$ (with $s \geq 1$). Then we have*

$$\log \mathbf{f}^{\underline{D}} = \underline{D} \log(\mathbf{f}),$$

where log *here means that we are taking the* log *coordinate by coordinate.*

Proof. It easily follows from direct computations: set $\underline{D} = (d_i^j)$; then $\mathbf{f}^{\underline{D}}$ is a s-uple $\mathbf{g} = (g_1, \ldots, g_s)^T$, with

$$g_i = \mathbf{f}^{\mathbf{d}_i} = \prod_{j=1}^{r} (f_i)^{d_i^j}.$$

Taking the log, we have

$$\log g_i = \sum_{j=1}^{r} d_i^j \log f_j,$$

and we are done. □

Proposition 3.1.10. *Let $\{\mathbf{f}_n\}_n$ be a sequence of r-uples of formal power series in $\mathbf{x}$, and let $\{\underline{D}_n\}_n$ be a sequence of matrices in $\mathcal{M}(s \times r, \mathbb{Q})$ (with $s \geq 1$). Then*

$$\prod_n (1 + \mathbf{f}_n)^{\underline{D}_n}$$

converges if and only if

$$\sum_n \underline{D}_n \mathbf{f}_n$$

does.

Proof. It follows from Lemma 3.1.9 and the analogous result in dimension one, taking the log of the absolute value. □

Definition 3.1.11. To simplify notations on the sums, we shall use the following encoding.

In Definition 3.1.2 and Remark 3.1.8 we saw three sums.

- The first one with $\mathbf{i} \in \mathcal{M}(1 \times d, \mathbb{N})$. In this case we shall simply write
$$\sum_{\mathbf{i}} .$$
- The second sum was with $I \in \mathcal{M}(n \times d, \mathbb{N})$. In this case we shall write
$$\sum_{I(n)} .$$
We shall call n the **dimension** of I.
- The last sum was with $\mathbf{I} = (I^1, \ldots, I^r)$, where $I^l \in \mathcal{M}(n^l \times d, \mathbb{N})$. Here we shall write
$$\sum_{\mathbf{I}(\mathbf{n})} .$$
We shall call $\mathbf{n}$ the **(pluri-)dimension** of $\mathbf{I}$.

In general, when writing some conditions for a sum, if we sum vectors of different dimensions, we shall mean that the smaller vector has 0 in the non-written variables.

As a convention, a multi-index of dimension 0 is an empty multi-index.

We shall often need to study the coefficients of a formal power series, that will depend on some data, and some variables (*e.g.*, r formal power series $\boldsymbol{\phi}(\mathbf{x})$). We will focus on terms that depend on the higher degree possible for our variables; we shall denote then by $\mathrm{l.o.t.}(\boldsymbol{\phi})$ all terms that depend on coefficients of $\boldsymbol{\phi}$ of degrees smaller than the ones we are considering, with respect to a certain total order on indices that shall be specified case by case.

To estimate the convergence of some formal power series, we shall need the following definition.

Definition 3.1.12. Let ϕ be a convergent power series in d complex variables $\mathbf{x} = (\mathbf{x}_1, \ldots, \mathbf{x}_d)^T$, and $\rho > 0$. The **majorant ρ-norm** of ϕ is

$$\square\phi\square_\rho := \sup\{|\phi(\mathbf{x})| \ : \ \|\mathbf{x}\| < \rho\} \in [0, +\infty].$$

We shall call

$$\mathcal{M}_\rho := \{\phi \ : \ \square\phi\square_\rho < +\infty\}$$

the **ρ-majorant space**.

3.1.2. Invariants

We shall consider the following (natural) generalization of rigid germs given in Definition 1.1.20.

Definition 3.1.13. Let $f : (\mathbb{C}^d, 0) \to (\mathbb{C}^d, 0)$ be a (dominant) holomorphic germ. We denote by $\mathcal{C}(f) = \{z \mid \det(df_z) = 0\}$ the **critical set** of f, and by $\mathcal{C}(f^\infty) = \bigcup_{n\in\mathbb{N}} f^{-n}\mathcal{C}(f)$ the **generalized critical set** of f. Then a (dominant) holomorphic germ f is **rigid** if:

(i) $\mathcal{C}(f^\infty)$ (is empty or) has simple normal crossings (SNC) at the origin; and
(ii) $\mathcal{C}(f^\infty)$ is forward f-invariant.

Especially in the last chapter, we shall be interested in another special class of holomorphic germs: strict germs.

Definition 3.1.14. Let $f : (\mathbb{C}^d, 0) \to (\mathbb{C}^d, 0)$ be a (dominant) holomorphic germ. Then f is a **strict germ** if there exist a SNC divisor with support C and a neighborhood U of 0 such that $f|_{U\setminus C}$ is a biholomorphism with its image.

In dimension strictly higher than 2, there are strict germs that are not rigid, as the following example shows.

Counterexample 3.1.15. Let us consider the holomorphic germ $f : (\mathbb{C}^3, 0) \to (\mathbb{C}^3, 0)$ given by

$$f(x, y, z) = \big(\lambda x, x(1 + y^2), y(1 + z)\big),$$

with $\lambda \neq 0$. If we compute the differential df, we get

$$df = \begin{pmatrix} \lambda & 0 & 0 \\ 1 + y^2 & 2xy & 0 \\ 0 & 1 + z & y \end{pmatrix},$$

and hence $\mathcal{C}(f) = \{xy = 0\}$. It follows by direct computation that $\mathcal{C}(f^\infty) = \mathcal{C}(f) = \{xy = 0\}$, but if we consider the image of $\{y = 0\}$, we get

$$f(x, 0, z) = (\lambda x, x, 0),$$

and hence $\mathcal{C}(f^\infty)$ is not (forward) f-invariant and f is not rigid.

On the other hand, we shall show that f is a biholomorphism with its image outside $\mathcal{C}(f^\infty)$. We want to show that every point $(X, Y, Z) \in f(xy \neq 0)$ has only one inverse (x, y, z) near the origin. From

$$(X, Y, Z) = f(x, y, z) = \big(\lambda x, x(1 + y^2), y(1 + z)\big)$$

we get

$$\begin{cases} x = \frac{X}{\lambda}, \\ y = \sqrt{\frac{\lambda Y}{X} - 1}, \\ z = \frac{Z}{y} - 1, \end{cases}$$

where y is actually multivalued: we have to decide which branch of $\sqrt{\cdot}$ we want to consider (there are no problems of branch points, since we have supposed $y \neq 0$). Let us denote by y_0 and $y_1 = -y_0$ the two possible values for y. Then, the point $z = Z/y - 1$ is near 0 if and only if $y \sim Z$; let us say that y_0 is such that $y_0 \sim Z$, then $y_1 \sim -Z$, and

$$z = \frac{Z}{y_1} - 1 \sim \frac{Z}{-Z} - 1 = -2,$$

that is far away from 0. It follows that for every point (X, Y, Z) in the image of $\{xy \neq 0\}$ there is exactly one preimage near the origin, and we are done.

Remark 3.1.16. Counterexample 3.1.15 can be easily generalized, with the same arguments, to germs of the form

$$f(x, y, z) = \big(g(x), a(x) + b(x)y^p, c(x)y^q z + d(x)y^k\big),$$

with $\lambda \in \mathbb{C}^*$, $p \geq 2, q, k \geq 1$, and a, b, c, d holomorphic functions in x not identically 0, and g an invertible holomorphic map.

Such an f is never rigid, but if $k \leq q$ and $\gcd(p, k) = 1$, then f is strict.

Here p, q, k play the same role of p, q, k in rigid germs of Class 4 of Theorem 2.4.3, and the condition that p, q, k have to satisfy in order to have a strict germ is the same as in [25, Proposition 1.5].

Definition 3.1.17. Let $f : (\mathbb{C}^d, 0) \to (\mathbb{C}^d, 0)$ be a rigid germ. Then f is called **q-reducible** if $\mathcal{C}(f^\infty)$ has q irreducible components.

Remark 3.1.18. The reducibility of a rigid germ is clearly a formal (and hence also holomorphic) invariant. Up to a change of coordinates one can suppose that a q-reducible rigid germ is such that $\mathcal{C}(f^\infty) = \{x_1 \cdot \ldots \cdot x_q = 0\}$.

Definition 3.1.19. Let $f : (\mathbb{C}^d, 0) \to (\mathbb{C}^d, 0)$ be a rigid germ. We shall call **total rank** of f the number of non-zero eigenvalues of df_0, or equivalently the rank of df_0^d.

Definition 3.1.20. Let $f : (\mathbb{C}^d, 0) \to (\mathbb{C}^d, 0)$ be a rigid germ. Let us consider Δ^d a small polydisc with center in 0. Being $\mathcal{C}(f^\infty)$ backward invariant, f induces a map $f_* : \pi_1(\Delta^d \setminus \mathcal{C}(f^\infty)) \to \pi_1(\Delta^d \setminus \mathcal{C}(f^\infty))$.

Remark 3.1.21. If $f : (\mathbb{C}^d, 0) \to (\mathbb{C}^d, 0)$ is a q-reducible rigid germ, then $\pi_1(\Delta^d \setminus \mathcal{C}(f^\infty)) \cong \mathbb{Z}^q$. So f_* is given by a $q \times q$ matrix $\underline{M} = \underline{M}(f)$ with integer entries.

On the other hand, if $\mathbf{x}$ are coordinates such that $\mathcal{C}(f^\infty) = \{x_1 \cdot \ldots \cdot x_q = 0\}$, then it follows directly from the condition of $\mathcal{C}(f^\infty)$ being f-backward invariant, that

$$\mathbf{f}(\mathbf{x}) = \left(\boldsymbol{\alpha}\mathbf{x}_1^{\underline{A}}\big(1 + \mathbf{g}(\mathbf{x})\big), \mathbf{h}(\mathbf{x})\right)^T, \tag{3.1}$$

where $\mathbf{x} = (x_1, \ldots, x_d)^T = (\mathbf{x}_1, x_{q+1}, \ldots, x_d)^T$, $\boldsymbol{\alpha} \in (\mathbb{C}^*)^q$, $\mathbf{g} : (\mathbb{C}^d, 0) \to \mathbb{C}^q$ and $\mathbf{h} : (\mathbb{C}^d, 0) \to \mathbb{C}^{d-q}$ are such that $\mathbf{g}(\mathbf{0}) = \mathbf{0}$ and $\mathbf{h}(\mathbf{0}) = \mathbf{0}$, and $\underline{A}$ is a suitable matrix in $\mathcal{M}(q \times q, \mathbb{N})$.

By direct computations, one can compute $\underline{M}$ for a germ as in (3.1), and find that

$$\underline{M} = \underline{A}.$$

Definition 3.1.22. Let $f : (\mathbb{C}^d, 0) \to (\mathbb{C}^d, 0)$ be a q-reducible rigid germ. We shall call $\underline{A} = \underline{A}(f) \in \mathcal{M}(q \times q, \mathbb{N})$ as in Remark 3.1.21 the **internal action** of f, and we shall say that f has **invertible internal action** if $\det \underline{A} \neq 0$, *i.e.*, if $\underline{A}$ is invertible in $\mathcal{M}(q \times q, \mathbb{Q})$. We shall call **internal rank** of f the number of non-zero eigenvalues of the linear part of $\mathbf{x} \mapsto \mathbf{x}^{\underline{A}}$, or equivalently the rank of the linear part of $\mathbf{x} \mapsto \mathbf{x}^{\underline{A}^q}$, where $\mathbf{x} = (x_1, \ldots, x_q)^T$.

Remark 3.1.23. Let $f : (\mathbb{C}^d, 0) \to (\mathbb{C}^d, 0)$ be a q-reducible rigid germ of internal rank r and total rank s. Then we obviously have that $0 \leq r \leq s$ and $r \leq q$.

If $\underline{A} = \underline{A}(f)$, then r is just the number of rows $\mathbf{a}_i = \mathbf{e}_i \underline{A}$ of $\underline{A}$ such that $\mathbf{e}_i \underline{A}^n = \mathbf{e}_i$ for a suitable $n \gg 0$. In particular, up to permuting coordinates, and up to replacing f by a suitable iterate, we can suppose

$$\underline{A} = \left(\begin{array}{c|c} \underline{I_r} & \underline{0} \\ \hline \underline{C} & \underline{D} \end{array} \right). \tag{3.2}$$

The general case would be with a $r \times r$ permutation matrix $\underline{P}$ replacing the role of $\underline{I_r}$. The condition of f being with invertible internal action translates on the fact that $\underline{D}$ is invertible.

Definition 3.1.24. Let $f : (\mathbb{C}^d, 0) \to (\mathbb{C}^d, 0)$ be a q-reducible attracting rigid germ with internal rank r as in (3.1), with $\underline{A} = \underline{A}(f)$ as in (3.2). Then we shall call $\underline{D} = \underline{D}(f) \in \mathcal{M}(p \times p, \mathbb{N})$ the **principal part** of f, and its dimension $p = q - r$ the **principal rank** of f.

The classification of attracting rigid germs in $\mathbb{C}^2$ given by Favre in [25] shows that class 7, the only class with a non-invertible internal action, can be studied by semi-conjugating the germ and obtaining a germ of class 4. With this idea in mind, and since the direct study of germs with an invertible internal action is easier, we shall start focusing on conjugacy classes for attracting rigid germs with an invertible internal action.

3.2. Classification

3.2.1. Poincaré-Dulac theory

The formal classification of attracting invertible germs is a classical result by Poincaré (1983) and Dulac (1904); the phenomenon that appears in this case is resonance.

Definition 3.2.1. Let $f : (\mathbb{C}^d, 0) \to (\mathbb{C}^d, 0)$ be an attracting germ, written in suitable coordinates $\mathbf{x}$ as

$$\mathbf{f}(\mathbf{x}) = \big(\boldsymbol{\lambda}\mathbf{x} + \mathbf{g}(\mathbf{x})\big),$$

where $\boldsymbol{\lambda} = (\lambda_1, \ldots, \lambda_d)^T$ is the vector of eigenvalues of df_0 (that we suppose in upper-triangular Jordan form). Then a monomial $\mathbf{x}^{\mathbf{n}}$ is called **resonant** for the k-th coordinate if either it arises in the k-th column of (the nilpotent part of) the Jordan form of $d\mathbf{f_0}$, or $\mathbf{n} = (n^1, \ldots, n^d)$ with $n^1 + \cdots + n^d \geq 2$ and

$$\boldsymbol{\lambda}^{\mathbf{n}} = \lambda_k.$$

Remark 3.2.2. In the literature, sometimes the monomials that arise in the Jordan form of the linear part of a germ f are not considered to be resonant.

The Poincaré-Dulac Theory will not tell us anything on the coordinates where we have a super-attracting behavior (in fact, if $\lambda_j = 0$, then every monomial is resonant for the j-th coordinate), but can gives us very useful informations on the linear part.

Proposition 3.2.3. *Let $f : (\mathbb{C}^d, 0) \to (\mathbb{C}^d, 0)$ be an attracting germ, whose differential df_0 has $\boldsymbol{\lambda} = (\lambda_1, \ldots, \lambda_d)$ as eigenvalues. Let us suppose that*

$$1 > |\lambda_1| \geq |\lambda_2| \geq \ldots \geq |\lambda_s| > |\lambda_{s+1}| = \ldots = |\lambda_d| = 0, \tag{3.3}$$

where $s \in \{0, \ldots, d\}$ is the total rank of f. Then f has only a finite number of resonant monomials for the k-th coordinate for every $k = 1, \ldots, s$; moreover a resonant monomial $\mathbf{x}^{\mathbf{n}}$ for the k-th coordinate is such that $\mathbf{n} = (n^1, \ldots, n^{k-1}, 0, \ldots, 0)$ for every $k = 1, \ldots, s$.

Proof. Let us consider a general resonant relation for the k-th coordinate:

$$\lambda_1^{n^1} \cdot \ldots \cdot \lambda_d^{n^d} = \lambda_k.$$

If $k \leq s$, then $n^j = 0$ for every $j > s$ (otherwise we would have $\lambda_k = 0$), and also for $j \geq k$, since if we would take the absolute value on both members, we would have that $\left|\boldsymbol{\lambda}^{\mathbf{m}}\right| \geq 1$ for a suitable $\mathbf{m} \in \mathbb{N}^d \setminus \{0\}$, that is in contrast with the assumption of attractivity of $\mathbf{f}$.

So we only have to deal with the case of an expression of the form

$$\lambda_1^{n^1} \ldots \lambda_j^{n^k} = \lambda_{k+1} =: \mu.$$

Taking the log of the absolute value (or equivalently Re log), we have

$$\sum_{i=1}^{k} n^i \operatorname{Re} \log \lambda_i = \operatorname{Re} \log \mu.$$

Being $\mathbf{f}$ attracting, then $\operatorname{Re}\log\lambda_i < 0$ for every i and $\operatorname{Re}\log\mu < 0$, and

$$n^i \leq \frac{-\operatorname{Re}\log\mu}{\min_k\{-\operatorname{Re}\log\lambda_k\}},$$

and we are done. □

Theorem 3.2.4 (Formal Poincaré-Dulac). *Let $f : (\mathbb{C}^d, 0) \to (\mathbb{C}^d, 0)$ be an attracting germ. Then, up to a formal change of coordinates, we have*

$$\mathbf{f} = \big(\boldsymbol{\lambda}\mathbf{x} + \mathbf{g}(\mathbf{x})\big),$$

where $\boldsymbol{\lambda} = (\lambda_1, \ldots, \lambda_d)^T$ is the vector of eigenvalues of df_0, and $\mathbf{g} = (g_1, \ldots, g_d)^T$ with g_k that contains only resonant monomials for the k-th coordinate.

Corollary 3.2.5. *Let $f : (\mathbb{C}^d, 0) \to (\mathbb{C}^d, 0)$ be an attracting germ. Then for every $N \in \mathbb{N}$ there exists a polynomial change of coordinates such that*

$$\mathbf{f} = \big(\boldsymbol{\lambda}\mathbf{x} + \mathbf{g}(\mathbf{x}) + \mathbf{R}(\mathbf{x})\big),$$

where $\boldsymbol{\lambda} = (\lambda_1, \ldots, \lambda_d)^T$ is the vector of eigenvalues of df_0, $\mathbf{g} = (g_1, \ldots, g_d)^T$ with g_k that contains only resonant monomials for the k-th coordinate, and $\mathbf{R} = (R_1, \ldots, R_d)^T$ with $R_k \in \mathfrak{m}^N$.

For analogous of the following result for invertible attracting germs, see [59] or [54]; we also refer to [6, Chapter 4] for an extensive exposition on this subject.

Theorem 3.2.6 (Poincaré-Dulac). *Let $f : (\mathbb{C}^d, 0) \to (\mathbb{C}^d, 0)$ be an attracting germ. Then, up to a holomorphic change of coordinates, we have*

$$\mathbf{f}(\mathbf{x}) = \big(\boldsymbol{\lambda}\mathbf{x} + \mathbf{g}(\mathbf{x})\big), \tag{3.4}$$

where $\boldsymbol{\lambda} = (\lambda_1, \ldots, \lambda_d)^T$ is the vector of eigenvalues of df_0, and $\mathbf{g} = (g_1, \ldots, g_d)^T$ with g_k that contains only resonant monomials for the k-th coordinate.

Proof. Thanks to Corollary 3.2.5, we can suppose that $\mathbf{f}$ is of the form

$$\mathbf{f}(\mathbf{x}) = \Big(\boldsymbol{\lambda}_1\mathbf{x}_1 + \boldsymbol{\delta}(\mathbf{x}) + \mathbf{R}(\mathbf{x}), \mathbf{h}(\mathbf{x})\Big)^T, \tag{3.5}$$

where $\boldsymbol{\lambda}_1 = (\lambda_1, \ldots, \lambda_s)^T$ is the vector of non-zero eigenvalues of df_0, ordered as in 3.3, $\boldsymbol{\delta} = (\delta_1, \ldots, \delta_s)^T$ is such that δ_k has only resonant monomials for the k-th coordinate, $\mathbf{R} = (R_1, \ldots, R_s)^T$ is such that $R_k \in$

$\mathfrak{m}^N$ for an arbitrarily big $N \in \mathbb{N}$, and $\mathbf{h} = (h_{s+1}, \ldots, h_d)^T$ are such that $h_k(\mathbf{x}) - \varepsilon_k x_{k-1} \in 2$ for every k, where ε_k arises from the nilpotent part of the Jordan form of df_0. Let us consider a candidate $\widetilde{\mathbf{f}}$ for a map conjugated to $\mathbf{f}$, of the form

$$\widetilde{\mathbf{f}}(\mathbf{x}) = \left(\boldsymbol{\lambda}_1\mathbf{x}_1 + \delta(\mathbf{x}) + \widetilde{\mathbf{R}}(\mathbf{x}), \widetilde{\mathbf{h}}(\mathbf{x})\right)^T,$$

with $\widetilde{\mathbf{R}}$ and $\widetilde{\mathbf{h}}$ analogous to $\mathbf{R}$ and $\mathbf{h}$ of (3.5). We shall prove that we can find a conjugation $\boldsymbol{\Phi}$ such that $\mathbf{f}$ is conjugated to $\widetilde{\mathbf{f}}$ with $\widetilde{\mathbf{R}} \equiv \mathbf{0}$ by induction on the coordinates. Let us suppose that $R_i \equiv 0$ for every $i = 1, \ldots, j-1$, and let us find a conjugation

$$\boldsymbol{\Phi}(\mathbf{x}) = \left(x_1, \ldots, x_{j-1}, x_j + \phi(\mathbf{x}), x_{j+1}, \ldots, x_d\right),$$

with $\phi \in \mathfrak{m}^2$. From the conjugacy relation $\boldsymbol{\Phi} \circ \mathbf{f} = \widetilde{\mathbf{f}} \circ \boldsymbol{\Phi}$ in the j-th coordinate, we find

$$\begin{aligned}(\boldsymbol{\Phi} \circ \mathbf{f})_j(\mathbf{x}) &= \lambda_j x_j + \delta_j(x_1, \ldots, x_{j-1}) + R_j(\mathbf{x}) + \phi \circ \mathbf{f}(\mathbf{x}) \\ (\widetilde{\mathbf{f}} \circ \boldsymbol{\Phi})_j &= \lambda_j x_j + \lambda_j \phi(\mathbf{x}) + \delta_j(x_1, \ldots, x_{j-1}),\end{aligned}$$

and hence

$$R_j(\mathbf{x}) + \phi \circ \mathbf{f}(\mathbf{x}) = \lambda_j \phi(\mathbf{x}).$$

Then a solution of this equation is

$$\phi(\mathbf{x}) = \sum_{n=0}^{\infty} \lambda_j^{-n-1} R_j \circ \mathbf{f}^{\circ n}(\mathbf{x}).$$

Since $R_j \in \mathfrak{m}^N$, then there exist $M > 0$ such that $\left|R_j(\mathbf{x})\right| \leq M \|\mathbf{x}\|^N$ for $\|\mathbf{x}\|$ small enough, while being $\mathbf{f}$ attracting, we have that $\left\|\mathbf{f}^{\circ n}(\mathbf{x})\right\| \leq \Lambda^n \|\mathbf{x}\|$ for $\left|\lambda_j\right| < \Lambda = (\left|\lambda_j\right| + 1)/2 < 1$. Then we get

$$\begin{aligned}|\phi(\mathbf{x})| &\leq \sum_{n=0}^{\infty} \left|\lambda_j\right|^{-n-1} M \Lambda^{nN} \|\mathbf{x}\|^N \\ &=\leq \sum_{n=0}^{\infty} \left|\lambda_j\right|^{-n-1} M \left|\lambda_j\right|^{-1} \left(\left|\lambda_j\right|^{-1} \Lambda^N\right)^n \|\mathbf{x}\|^N,\end{aligned}$$

that converges for $\|\mathbf{x}\|$ small enough if we pick N such that $\Lambda^N < \lambda_j$. □

Definition 3.2.7. An attracting germ $f : (\mathbb{C}^d, 0) \to (\mathbb{C}^d, 0)$ written in suitable coordinates as

$$\mathbf{f}(\mathbf{x}) = \big(\boldsymbol{\lambda}\mathbf{x} + \mathbf{g}(\mathbf{x})\big), \tag{3.4}$$

where $\boldsymbol{\lambda} = (\lambda_1, \ldots, \lambda_d)^T$ is the vector of eigenvalues of df_0, and $\mathbf{g} = (g_1, \ldots, g_d)^T$ with g_k that contains only resonant monomials for the k-th coordinate, is called in **Poincaré-Dulac normal form**.

Remark 3.2.8. In particular, this result shows how we can choose a polynomial normal form for invertible germs. While this thing is still true for all attracting rigid germs in dimension 2 (see Theorem 2.4.3), we shall see that it is no longer true in higher dimensions (see Remark 3.2.25).

3.2.2. Topological Resonances

Remark 3.2.9. We want now to apply Theorem 3.2.6 to an attracting rigid germ of the form (3.1). The coordinates provided by Theorem 3.2.6 need not to be compatible with the monomial part of (3.1). This problem can be avoided by taking a suitable iterate of f (see Remark 3.1.23), so that the matrix $\underline{A}$ is of the form of (3.2).

In this case, we deduce that if we have $f : (\mathbb{C}^d, 0) \to (\mathbb{C}^d, 0)$ an attracting q-reducible rigid germ of total rank s and principal rank p, up to taking a suitable iterate, and up to change of coordinates, we can write it in the following way:

$$\mathbf{f}(\mathbf{x}) = \Big(\boldsymbol{\lambda}\mathbf{x}_1 + \boldsymbol{\delta}(\mathbf{x}_1), \boldsymbol{\alpha}\mathbf{y}^{\underline{B}}\big(1 + \mathbf{g}(\mathbf{x})\big), \mathbf{h}(\mathbf{x})\Big)^T, \tag{3.6}$$

where $\mathbf{x} = (x_1, \ldots, x_d)^T$, $\mathbf{x}_1 = (x_1, \ldots, x_s)^T$ and $\mathbf{y} = (\mathbf{x}_1, x_{s+1}, \ldots, x_{s+p})^T$; $\boldsymbol{\delta} = (\delta_1, \ldots, \delta_s)^T$, $\delta_k \in \mathfrak{m}^2$ has only resonant monomials for the k-th coordinate, $\mathbf{g} = (g_{s+1}, \ldots, g_{s+p})^T$ with $g_k \in \mathfrak{m}$ for every k, and $\mathbf{h} = (h_{s+p+1}, \ldots, h_d)^T$ with $h_k - \varepsilon_k x_{k-1} \in \mathfrak{m}^2$ for every k, for a suitable $\varepsilon_k \in \{0, 1\}$ to take care of possible contributions to the nilpotent part of df_0.. Here $\boldsymbol{\lambda} = (\lambda_1, \ldots, \lambda_s)^T$ is a s-uple of the non-zero eigenvalues of $d\mathbf{f_0}$, so we have $0 < |\lambda_i| < 1$ for $i = 1, \ldots, s$, being f attracting.

As in (3.2), we split the matrix $\underline{B}$ as

$$\underline{B} := \big(\, \underline{C} \mid \underline{D} \,\big). \tag{3.7}$$

For the general case, one can adapt the resonance condition of Definition 3.2.1 to take care of the fact that the linear part of f is not in Jordan normal form, and deduce an analogous of Theorem 3.2.6.

Starting then from a rigid germ as in (3.6), and hence in particular in Poincaré-Dulac normal form, we would like now to change coordinates in order to simplify the second (bunch of) coordinates, but still obtaining a germ in Poincaré-Dulac normal form. This process is called "renormalization", and for rigid germs in higher dimensions, it arises a new phenomenon that does not appear in dimension 2: topological resonances.

Definition 3.2.10. Let $f : (\mathbb{C}^d, 0) \to (\mathbb{C}^d, 0)$ be an attracting q-reducible rigid germ of total rank s and principal rank p, of the form (3.6). Then a monomial $\mathbf{x}_1^{\mathbf{n}}$ with $\mathbf{n} \in \mathcal{M}(1 \times s, \mathbb{N})$ and $\mathbf{x}_1 = (x_1, \dots, x_s)^T$ is called **topologically resonant** for $\mathbf{f}$ if $\boldsymbol{\lambda}^{\mathbf{n}}$ is an eigenvalue for $\underline{D}$.

Proposition 3.2.11. *Let $f : (\mathbb{C}^d, 0) \to (\mathbb{C}^d, 0)$ be an attracting rigid germ of the form* (3.6). *Then f has only a finite number of topologically resonant monomials.*

Proof. Let us set $\boldsymbol{\lambda}$ and $\underline{D} = \underline{D}(f)$ as in Definition 3.2.10.

First of all, $\underline{D}$ has only a finite number of eigenvalues μ.

Let us suppose that there exists $\mathbf{n}$ such that $\boldsymbol{\lambda}^{\mathbf{n}} = \mu$, i,e,,

$$\lambda_1^{n^1} \dots \lambda_s^{n^s} = \mu.$$

Taking the log of the absolute value (or equivalently Re log), we have

$$\sum_{i=1}^{s} n^i \operatorname{Re}\log \lambda_i = \operatorname{Re}\log \mu.$$

Being $\mathbf{f}$ attracting, then $\operatorname{Re}\log \lambda_i < 0$ for every i, then we must have $\operatorname{Re}\log \mu < 0$, and

$$n^i \le \frac{-\operatorname{Re}\log \mu}{\min_k\{-\operatorname{Re}\log \lambda_k\}},$$

and we are done. □

Theorem 3.2.12. *Let $f : (\mathbb{C}^d, 0) \to (\mathbb{C}^d, 0)$ be an attracting q-reducible rigid germ with invertible internal action, of total rank s and principal rank p. Up to taking iterates, and up to (formal) change of coordinates, we have*

$$\mathbf{f}(\mathbf{x}) = \Big(\boldsymbol{\lambda}\mathbf{x}_1 + \boldsymbol{\delta}(\mathbf{x}_1), \boldsymbol{\alpha}\mathbf{y}^{\underline{B}}(1 + \mathbf{g}(\mathbf{x}_1)), \mathbf{h}(\mathbf{x})\Big)^T, \qquad (3.16)$$

with the same conditions as in Remark 3.2.9, *where* $\mathbf{g}$ *contains only topologically resonant monomials.*

Proof. Let us take $\mathbf{f}$ as in (3.6) and $\widetilde{\mathbf{f}}$ as in (3.16), but with $\widetilde{\boldsymbol{\delta}}$, $\widetilde{\mathbf{g}}$ and $\widetilde{\mathbf{h}}$ instead of $\boldsymbol{\delta}$, $\mathbf{g}$ and $\mathbf{h}$ respectively; then let us consider a conjugation of the form

$$\boldsymbol{\Phi}(\mathbf{x}) = \Big(\mathbf{x}_1, \mathbf{x}_2\big(1+\boldsymbol{\phi}(\mathbf{x})\big), \mathbf{x}_3\Big)^T,$$

with $\mathbf{x} = (\mathbf{x}_1, \mathbf{x}_2, \mathbf{x}_3)^T$.

Taking the second group of coordinates for the conjugacy relation

$$\boldsymbol{\Phi} \circ \mathbf{f} = \widetilde{\mathbf{f}} \circ \boldsymbol{\Phi},$$

and simplifying the factor $\boldsymbol{\alpha}\mathbf{y}^{\underline{B}}$, we get

$$\begin{aligned} &\mathbb{I} := \big(1+\mathbf{g}(\mathbf{x})\big)\big(1+\boldsymbol{\phi}\circ\mathbf{f}(\mathbf{x})\big) \\ &|| \\ &\mathbb{II} := \big(1+\boldsymbol{\phi}(\mathbf{x})\big)^{\underline{D}}\big(1+\widetilde{\mathbf{g}}\circ\boldsymbol{\Phi}(\mathbf{x})\big). \end{aligned} \tag{3.8}$$

Let us set $1+\phi_k(\mathbf{x}) = \sum_{\mathbf{i}} \phi_{k,\mathbf{i}}\mathbf{x}^{\mathbf{i}}$, then $1+g_k(\mathbf{x}) = \sum_{\mathbf{i}} g_{k,\mathbf{i}}\mathbf{x}^{\mathbf{i}}$ and analogously for $1+\widetilde{g}_k$ for all $k = s+1, \ldots, s+p$. Moreover, set $\lambda_k x_k + \delta_k(\mathbf{x}) = \sum_{\mathbf{i}} \delta_{k,\mathbf{i}}\mathbf{x}^{\mathbf{i}}$ for all $k = 1, \ldots, s$ and finally $h_k(\mathbf{x}) = \sum_{\mathbf{i}} h_{k,\mathbf{i}}\mathbf{x}^{\mathbf{i}}$ for all $k = s+p+1, \ldots, d$.

Then

$$\begin{aligned} \mathbb{I}_k &= \sum_{\mathbf{i}} \phi_{k,\mathbf{i}}\big(\boldsymbol{\lambda}\mathbf{x}_1 + \boldsymbol{\delta}(\mathbf{x}_1)\big)^{\mathbf{i}_1}\boldsymbol{\alpha}^{\mathbf{i}_2}\mathbf{y}^{\mathbf{i}_2\underline{B}}\big(1+\mathbf{g}(\mathbf{x})\big)^{\mathbf{i}_2+\mathbf{e}_k}\big(\mathbf{h}(\mathbf{x})\big)^{\mathbf{i}_3}. \\ &= \sum_{\mathbf{i}} \phi_{k,\mathbf{i}}\boldsymbol{\alpha}^{\mathbf{i}_2}\mathbf{y}^{\mathbf{i}_2\underline{B}} \sum_{\mathbf{I}(\mathbf{i}_1)} \delta_{\mathbf{I}}\mathbf{x}^{|\mathbf{I}|} \sum_{\mathbf{J}(\mathbf{i}_2+\mathbf{e}_k)} g_{\mathbf{J}}\mathbf{x}^{|\mathbf{J}|} \sum_{\mathbf{L}(\mathbf{i}_3)} h_{\mathbf{L}}\mathbf{x}^{|\mathbf{L}|}, \end{aligned}$$

where we split $\mathbf{i} = (\mathbf{i}_1, \mathbf{i}_2, \mathbf{i}_3)$.

If we denote by $\mathbb{I}_{k,\mathbf{n}}$ the coefficient of $\mathbf{x}^{\mathbf{n}}$ of $\mathbb{I}_k$, *i.e.*,

$$\mathbb{I}_k = \sum_{\mathbf{n}} \mathbb{I}_{k,\mathbf{n}}\mathbf{x}^{\mathbf{n}},$$

then we have

$$\mathbb{I}_{k,\mathbf{n}} = \sum_{\substack{\mathbf{i};\mathbf{I}(\mathbf{i}_1);\mathbf{J}(\mathbf{i}_2+\mathbf{e}_k);\mathbf{L}(\mathbf{i}_3) \\ \mathbf{i}_2\underline{B}+|\mathbf{I}|+|\mathbf{J}|+|\mathbf{L}|=\mathbf{n}}} \phi_{k,\mathbf{i}}\boldsymbol{\alpha}^{\mathbf{i}_2}\delta_{\mathbf{I}}g_{\mathbf{J}}h_{\mathbf{L}}. \tag{3.9}$$

Analogously

$$\begin{aligned} \mathbb{II}_k &= \sum_{\mathbf{j}} \widetilde{g}_{k,\mathbf{j}}\mathbf{x}^{\mathbf{j}}\big(1+\boldsymbol{\phi}(\mathbf{x})\big)^{\mathbf{e}_k\underline{D}+\mathbf{j}_2} \\ &= \sum_{\mathbf{j}} \widetilde{g}_{k,\mathbf{j}}\mathbf{x}^{\mathbf{j}} \sum_{\mathbf{K}(\mathbf{e}_k\underline{D}+\mathbf{j}_2)} \phi_{\mathbf{K}}\mathbf{x}^{|\mathbf{K}|}. \end{aligned}$$

Let us denote by $\mathbb{III}_{k,\mathbf{n}}$ the coefficient of $\mathbf{x}^{\mathbf{n}}$ of $\mathbb{III}_k$; then we have

$$\mathbb{III}_{k,\mathbf{n}} = \sum_{\substack{\mathbf{j}:\mathbf{K}(\mathbf{e}_k\underline{D}+\mathbf{j}_2) \\ \mathbf{j}+|\mathbf{K}|=\mathbf{n}}} \widetilde{g}_{k,\mathbf{j}}\phi_{\mathbf{K}}. \tag{3.10}$$

We want now to study the higher order terms in $\mathbb{I}_{k,\mathbf{n}}$ and $\mathbb{III}_{k,\mathbf{n}}$ with respect to the following order: for an index $\mathbf{m} = (m^1, \ldots, m^d)$, we first consider the value $m^1 + \ldots + m^d$, and then the lexicographic order on all the variables.

Then we have

$$\mathbb{III}_{k,\mathbf{n}} = \mathbf{e}_k\underline{D}\boldsymbol{\phi}_{\mathbf{n}} + \mathrm{l.\,o.\,t.}(\boldsymbol{\phi}), \tag{3.11}$$

where $\boldsymbol{\phi}_{\mathbf{n}} = \left(\phi_{s+1,\mathbf{n}}, \ldots, \phi_{s+p,\mathbf{n}}\right)^T$.

In fact, from (3.10), to have the highest order for $\boldsymbol{\phi}$, we must take $\mathbf{j} = \mathbf{0}$; then K has pluri-dimension $\mathbf{e}_k\underline{D}$; to have the highest order, we should maximize the order of one of the factors of $\phi_{\mathbf{K}}$, and take the order 0 for the others. Then (3.11) follows.

For (3.9) we are now interested in the pluri-order $\mathbf{n}$: we get

$$\mathbb{I}_{k,\mathbf{n}} = \boldsymbol{\lambda}^{\mathbf{n}_1}\phi_{k,\mathbf{n}_1}\mathbf{1}^{\mathbf{0}}_{\mathbf{n}_2}\mathbf{1}^{\mathbf{0}}_{\mathbf{n}_3} + \mathrm{l.\,o.\,t.}(\boldsymbol{\phi}), \tag{3.12}$$

where $\mathbf{1}$ denotes the Kronecker's delta function.

Indeed, from (3.9), since

$$\mathbf{i}_2\underline{B} = \left(\, \mathbf{i}_2\underline{C} \, \middle| \, \mathbf{i}_2\underline{D} \, \right),$$

and $\underline{D}$ is such that $\mathbf{e}_k\underline{D} > \mathbf{e}_k$, we have $\mathbf{i}_2\underline{D} > \mathbf{i}_2$ if $\mathbf{i}_2 \neq \mathbf{0}$, and hence we must have $\mathbf{n}_2 = \mathbf{0}$. For analogous reasons we have $\mathbf{n}_3 = \mathbf{0}$, and (3.12) follows because the linear part of $\boldsymbol{\delta}$ is upper triangular, and thanks to the chosen order on indices.

Putting together all coordinates, from

$$\mathbb{I}_{\mathbf{n}} = \mathbb{III}_{\mathbf{n}}, \tag{3.13}$$

we get

$$\underline{D}\boldsymbol{\phi}_{\mathbf{n}} = \boldsymbol{\lambda}^{\mathbf{n}_1}\boldsymbol{\phi}_{\mathbf{n}_1}\mathbf{1}^{\mathbf{0}}_{\mathbf{n}_2}\mathbf{1}^{\mathbf{0}}_{\mathbf{n}_3} + \mathrm{l.\,o.\,t.}(\boldsymbol{\phi}). \tag{3.14}$$

So if $\mathbf{n}$ is not resonant, we can define $\boldsymbol{\phi}_{\mathbf{n}}$ recursively, and take $\widetilde{g}_{k,\mathbf{n}} = 0$ for each k.

If $\mathbf{n}$ is resonant, we first notice that

$$\mathbb{III}_{k,\mathbf{n}} = \widetilde{g}_{k,\mathbf{n}} + \mathrm{l.\,o.\,t.}(\widetilde{\mathbf{g}}). \tag{3.15}$$

Then we can define (recursively) $\widetilde{\mathbf{g}}_{\mathbf{n}}$ such that (3.13) holds even for resonant monomials. $\square$

Remark 3.2.13. We could be more precise in Theorem 3.2.12. In fact, if $\mathbf{n}$ is resonant then equation (3.14) is a linear system, represented by $\underline{L} = \underline{D} - \boldsymbol{\lambda}^{\mathbf{n}}\underline{I}_p$, on $\boldsymbol{\phi}_{\mathbf{n}}$ (which has p variables and p equations) that is not invertible, but it has a kernel of dimension equal to the geometric multiplicity $m_g = \dim \mathrm{Ker}(\underline{D} - \boldsymbol{\lambda}^{\mathbf{n}}\underline{I}_p)$.

Let us suppose that the first $h := p - m_g$ lines of $\underline{L}$ are linearly independent: then the last m_g lines of $\underline{L}$ gives us, thanks to (3.15), m_g linear conditions on $\widetilde{g}_{\mathbf{n}}$ to be satisfied, while from the first h lines of $\underline{L}$ we can set h coordinates of $\boldsymbol{\phi}_{\mathbf{n}}$ to be zero.

Remark 3.2.14. We have seen how we can define recursively the conjugation Φ. In particular, we can obtain the "normal" form (3.16) up to an arbitrarily high order M by conjugating by a polynomial map (and hence, a biholomorphism).

Theorem 3.2.15. *Let $f : (\mathbb{C}^d, 0) \to (\mathbb{C}^d, 0)$ be an attracting rigid germ with invertible internal action of total rank s. Up to taking iterates, and up to a* holomorphic *change of coordinates, we have*

$$\mathbf{f}(\mathbf{x}) = \Big(\boldsymbol{\lambda}\mathbf{x}_1 + \boldsymbol{\delta}(\mathbf{x}_1), \boldsymbol{\alpha}\mathbf{y}^{\underline{B}}\big(1 + \mathbf{g}(\mathbf{x}_1)\big), \mathbf{h}(\mathbf{x})\Big)^T, \tag{3.16}$$

with the same conditions as in Remark 3.2.9, where $\mathbf{g}$ *contains only topologically resonant monomials.*

Proof. Let us use the same notations as in the proof of Theorem 3.2.12. Recalling Remark 3.2.14, we can suppose that $\mathbf{f}$ is as in (3.6), with

$$\mathbf{g}(\mathbf{x}) = \widetilde{\mathbf{g}}(\mathbf{x}_1) + \mathbf{R}(\mathbf{x}),$$

with $\mathbf{R} = (R_{s+1}, \ldots, R_{s+p})^T$ and $R_k \in \mathfrak{m}^M$ for every k, for a suitable M to be chosen. Then (3.8) in this case is

$$\big(1 + \widetilde{\mathbf{g}}(\mathbf{x}_1) + \mathbf{R}(\mathbf{x})\big)\big(1 + \boldsymbol{\phi} \circ \mathbf{f}(\mathbf{x})\big) = \big(1 + \boldsymbol{\phi}(\mathbf{x})\big)^{\underline{D}}\big(1 + \widetilde{\mathbf{g}}(\mathbf{x}_1)\big), \tag{3.17}$$

since $\widetilde{\mathbf{g}}$ does not depend on $\mathbf{x}_2$.

But in this case, we can find a direct solution of (3.17), given by

$$1 + \boldsymbol{\phi} = \prod_{k=1}^{\infty} \left(1 + \mathbf{e} \circ \mathbf{f}^{\circ k-1}\right)^{\underline{D}^{-k}}, \tag{3.18}$$

where

$$\mathbf{e}(\mathbf{x}) = \frac{\mathbf{R}(\mathbf{x})}{1 + \widetilde{\mathbf{g}}(\mathbf{x}_1)}.$$

We notice that each component of $\mathbf{e}$ is still in $\mathfrak{m}^M$.

If we set

$$\boldsymbol{\phi}_n(\mathbf{x}) = \prod_{k=1}^{n} \left(1 + \mathbf{e} \circ \mathbf{f}^{\circ k-1}\right)^{\underline{D}^{-k}},$$

then we have

$$\begin{aligned}\left(1 + \mathbf{e}(\mathbf{x})\right)\left(1 + \boldsymbol{\phi}_n \circ \mathbf{f}(\mathbf{x})\right) &= \left(1 + \mathbf{e}(\mathbf{x})\right) \prod_{k=1}^{n} \left(1 + \mathbf{e} \circ \mathbf{f}^{\circ k}(\mathbf{x})\right)^{\underline{D}^{-k}} \\ &= \prod_{k=1}^{n+1} \left(1 + \mathbf{e} \circ \mathbf{f}^{\circ k-1}(\mathbf{x})\right)^{\underline{D}^{-k+1}} \\ &= \left(1 + \boldsymbol{\phi}_{n+1}(\mathbf{x})\right)^{\underline{D}}.\end{aligned}$$

We only have to prove that $\boldsymbol{\phi}_n$ converges to $\boldsymbol{\phi} = \boldsymbol{\phi}_\infty$.

Thanks to Proposition 3.1.10, we just have to prove that

$$\sum_{k=1}^{\infty} \underline{D}^{-k} \left(\mathbf{e} \circ \mathbf{f}^{\circ k-1}\right)$$

converges in a neighborhood of 0. Since $e_k \in \mathfrak{m}^M$ for each k, we have that there exists $K > 0$ such that $\|\mathbf{e}(\mathbf{x})\| \leq K \|\mathbf{x}\|^M$, while being f attracting, there exists $0 < \Lambda < 1$ such that, for $\|\mathbf{x}\|$ small enough, we have $\left\|\mathbf{f}^{\circ k}(\mathbf{x})\right\| \leq \Lambda^k \|\mathbf{x}\|$.

Then we have for $\|\mathbf{x}\|$ small enough

$$\begin{aligned}\left\| \sum_{k=1}^{\infty} \underline{D}^{-k} \left(\mathbf{e} \circ \mathbf{f}^{\circ k-1}\right) \right\| &\leq \sum_{k=1}^{\infty} \left|\underline{D}^{-k}\right| \left\|\mathbf{e} \circ \mathbf{f}^{\circ k-1}\right\| \\ &\leq \sum_{k=1}^{\infty} \left|\underline{D}^{-1}\right|^k K \Lambda^{(k-1)M} \|\mathbf{x}\|^M.\end{aligned}$$

If we choose M big enough to have $\left|\underline{D}^{-1}\right| \Lambda^M < 1$, the sum converges and we are done. □

Remark 3.2.16. The main difference between the result of Theorem 3.2.15 and (3.6) is that in (3.16) $\mathbf{g}$ depends only on the first s coordinates. In particular, the coordinates f_k of $\mathbf{f}$ with $k = s+1, \ldots, s+p$ depends only on the first $s + p$ coordinates. This allows to construct for example an f-invariant foliation of codimension $s + p$, induced by $dx_1 \wedge \ldots \wedge dx_{s+p}$, that is actually a subfoliation of the f-invariant foliation of codimension s induced by $dx_1 \wedge \ldots \wedge dx_s$.

3.2.3. Affine actions

Remark 3.2.17. Let us suppose that $f : (\mathbb{C}^d, 0) \to (\mathbb{C}^d, 0)$ is an attracting $(d-1)$-reducible rigid germ with invertible internal action. Assume that the internal action is of the form (3.2) (always possible up to iterates). Let us denote by r the internal rank, s the total rank and p the principal rank of f. Then, thanks to Theorem 3.2.15, up to holomorphic conjugacy we can suppose that

$$\mathbf{f}(\mathbf{x}) = \Big(\boldsymbol{\lambda}\mathbf{x}_1, \boldsymbol{\alpha}\mathbf{y}^{\underline{B}}(1+\mathbf{g}(\mathbf{x}_1)), h(\mathbf{x})\Big)^T, \tag{3.19}$$

with $\mathbf{x} = (\mathbf{y}, x_d)^T, \mathbf{y} = (\mathbf{x}_1, \mathbf{x}_2)^T, \mathbf{x}_1 = (x_1, \dots, x_r)^T$, $\mathbf{g}$ with only resonant monomials. Moreover, we already dealt with the case of $\frac{\partial h}{\partial x_d}(\mathbf{0}) \neq 0$, or equivalently $s = r+1$ (see Theorem 3.2.4 (Poincaré-Dulac)), so we can suppose that $\frac{\partial h}{\partial x_d}(\mathbf{0}) = 0$.

Taking the differential df_0, we get

$$d\mathbf{f_0} = \left(\begin{array}{c|c|c} \boldsymbol{\lambda} : \underline{I} & \underline{0} & \mathbf{0} \\ \hline * & \underline{D} : \boldsymbol{\alpha}\mathbf{y}^{\underline{B}}(1+\mathbf{g}(\mathbf{x}_1)) \cdot\cdot\, \mathbf{x}_2^{-\underline{I}} & \mathbf{0} \\ \hline * & * & \frac{\partial h}{\partial x_d} \end{array}\right).$$

Since $\mathcal{C}(f^\infty) = \{x_1 \cdot \ldots \cdot x_{d-1} = 0\}$, we get

$$\frac{\partial h}{\partial x_d} = \mathbf{y}^{\mathbf{l}} u(\mathbf{x}),$$

with $\mathbf{l} \in \mathcal{M}(1 \times d-1, \mathbb{N})$ and $u(\mathbf{0}) \neq 0$. Integrating, we obtain

$$h(\mathbf{x}) = \mu \mathbf{y}^{\mathbf{l}}\big(1 + \varepsilon(\mathbf{x})\big) + p(\mathbf{y}),$$

with $\varepsilon(\mathbf{0}) = 0$ and $p(\mathbf{0}) = 0$ (and $\mathbf{l} \neq \mathbf{0}$, thanks to the condition on h). Summing up, we get

$$\mathbf{f}(\mathbf{x}) = \Big(\boldsymbol{\lambda}\mathbf{x}_1, \boldsymbol{\alpha}\mathbf{y}^{\underline{B}}(1+\mathbf{g}(\mathbf{x}_1)), \mu\mathbf{y}^{\mathbf{l}}\big(1+\varepsilon(\mathbf{x})\big) + p(\mathbf{y})\Big). \tag{3.20}$$

In this case, we can get the following result, an analogous of what happens in the 2-dimensional case for Classes 2 and 4 of Theorem 2.4.3 (see [25, pp. 491–494]).

Theorem 3.2.18. *Let $f : (\mathbb{C}^d, 0) \to (\mathbb{C}^d, 0)$ be an attracting $(d-1)$-reducible rigid germ with invertible internal action, of internal and total rank r. Up to taking iterates, and up to holomorphic conjugacy, we have*

$$\mathbf{f}(\mathbf{x}) = \Big(\boldsymbol{\lambda}\mathbf{x}_1, \boldsymbol{\alpha}\mathbf{y}^{\underline{B}}(1+\mathbf{g}(\mathbf{x}_1)), \mu\mathbf{y}^{\mathbf{l}}x_d + p(\mathbf{y})\Big)^T, \tag{3.21}$$

with $\boldsymbol{\lambda} = (\lambda_1, \ldots, \lambda_s)$ *is the vector of non-zero eigenvalues of* df_0, $\mathbf{x} = (\mathbf{y}, x_d)^T$, $\mathbf{y} = (\mathbf{x}_1, x_{r+1}, \ldots, x_{d-1})^T$, $\mathbf{l} \neq \mathbf{0} \in \mathcal{M}(1 \times d-1, \mathbb{N})$, $\mathbf{g}$ *with only (internal) resonant monomials, and* $p(\mathbf{y}) \in \mathfrak{m}$ *is a suitable analytic function.*

Proof. During the proof, we shall omit all the transpositions of vectors to have them vertical.

Thanks to Remark 3.2.17, we can suppose that $\mathbf{f}$ is of the form (3.20). Let us denote by $\widetilde{\mathbf{f}}$ the map we want to obtain, as in (3.21), with $\widetilde{p}$ instead of p.

We shall consider a conjugation of the form

$$\boldsymbol{\Phi}(\mathbf{x}) = \big(\mathbf{y}, x_d(1 + \phi(\mathbf{x}))\big),$$

and we will consider the conjugacy relation $\boldsymbol{\Phi} \circ \mathbf{f} = \widetilde{\mathbf{f}} \circ \boldsymbol{\Phi}$. Comparing the last coordinate of the conjugacy relation, we get:

$$\begin{aligned}
\big(\widetilde{\mathbf{f}} \circ \boldsymbol{\Phi}\big)_d(\mathbf{x}) &= \mu \mathbf{y}^{\mathbf{l}} x_d\big(1 + \phi(\mathbf{x})\big) + \widetilde{p}(\mathbf{y}) \\
&\,|| \qquad\qquad\qquad\qquad\qquad\qquad\qquad\qquad\qquad\qquad (3.22)\\
\big(\boldsymbol{\Phi} \circ \mathbf{f}\big)_d(\mathbf{x}) &= \mu \mathbf{y}^{\mathbf{l}} x_d\big(1 + \varepsilon(\mathbf{x})\big)\big(1 + \phi \circ \mathbf{f}(\mathbf{x})\big) + p(\mathbf{y})\big(1 + \phi \circ \mathbf{f}(\mathbf{x})\big). \qquad (3.23)
\end{aligned}$$

We want now to split (3.23) in two parts, one that depends on x_d, and one that does not.

Since

$$\int_0^1 \frac{d}{dt} \phi\big(\mathbf{f}(\mathbf{y}, t x_d)\big) dt = \phi\big(\mathbf{f}(\mathbf{x})\big) - \phi\big(\mathbf{f}(\mathbf{y}, 0)\big),$$

we get

$$\phi\big(\mathbf{f}(\mathbf{x})\big) = \phi\Big(\boldsymbol{\lambda}\mathbf{x}_1, \boldsymbol{\alpha}\mathbf{y}^{\underline{B}}(1 + \mathbf{g}(\mathbf{x}_1)), p(\mathbf{y})\Big) + \int_0^1 \frac{d}{dt} \phi\big(\mathbf{f}(\mathbf{y}, t x_d)\big) dt,$$

and

$$\begin{aligned}
\int_0^1 \frac{d}{dt} \phi\big(\mathbf{f}(\mathbf{y}, t x_d)\big) dt &= \int_0^1 \frac{\partial \phi}{\partial x_d}\big(\mathbf{f}(\mathbf{y}, t x_d)\big) \frac{\partial\big(\mu \mathbf{y}^{\mathbf{l}} x_d(1 + \varepsilon(\mathbf{x})) + p(\mathbf{y})\big)}{\partial x_d} x_d dt \\
&= \mu \mathbf{y}^{\mathbf{l}} x_d \int_0^1 \frac{\partial \phi}{\partial x_d}\big(\mathbf{f}(\mathbf{y}, t x_d)\big)\big(1 + \eta(\mathbf{y}, t x_d)\big) dt,
\end{aligned}$$

where $\eta(\mathbf{x}) := \varepsilon(\mathbf{x}) + x_d \frac{\partial \varepsilon}{\partial x_d}(\mathbf{x})$.

So (3.23) can be written as

$$\mu\mathbf{y}^{\mathbf{l}}x_d\Bigg(\big(1+\varepsilon(\mathbf{x})\big)\big(1+\phi\circ\mathbf{f}(\mathbf{x})\big) + p(\mathbf{y})\int_0^1\frac{\partial\phi}{\partial x_d}\big(\mathbf{f}(\mathbf{y},tx_d)\big)\big(1+\eta(\mathbf{y},tx_d)\big)dt\Bigg) + p(\mathbf{y})\left(1+\phi\Big(\boldsymbol{\lambda}\mathbf{x}_1,\boldsymbol{\alpha}\mathbf{y}^{\underline{B}}(1+\mathbf{g}(\mathbf{x}_1)),p(\mathbf{y})\Big)\right).$$

Then thanks to (3.22), we get

$$1+\phi(\mathbf{x}) = 1+\varepsilon(\mathbf{x})+T\phi(\mathbf{x}) \tag{3.24}$$

$$\widetilde{p}(\mathbf{y}) = p(\mathbf{y})\left(1+\phi\Big(\boldsymbol{\lambda}\mathbf{x}_1,\boldsymbol{\alpha}\mathbf{y}^{\underline{B}}(1+\mathbf{g}(\mathbf{x}_1)),p(\mathbf{y})\Big)\right), \tag{3.25}$$

where

$$T\phi(\mathbf{x}) := \big(1+\varepsilon(\mathbf{x})\big)\phi\circ\mathbf{f}(\mathbf{x}) + p(\mathbf{y})\int_0^1\frac{\partial\phi}{\partial x_d}\big(\mathbf{f}(\mathbf{y},tx_d)\big)\big(1+\eta(\mathbf{y},tx_d)\big)dt. \tag{3.26}$$

The candidate solution for (3.24) is given by

$$\phi = \sum_{k=0}^{\infty} T^k\varepsilon; \tag{3.27}$$

we need to prove that this sum is convergent.

Lemma 3.2.19 (Cauchy's Lemma). *For every* $0<\theta<1$ *and* $\phi\in\mathcal{M}_\rho$, *we have*

$$\max\left\{\left\|\frac{\partial\phi}{\partial x_1}\right\|_{\theta\rho},\ldots,\left\|\frac{\partial\phi}{\partial x_d}\right\|_{\theta\rho}\right\} \le \frac{\|\phi\|_\rho}{\rho(1-\theta)}.$$

If moreover $\phi(\mathbf{0})=0$, *then for every* $\mathbf{x}$ *such that* $\|\mathbf{x}\|\le\theta\rho$, *we have*

$$|\phi(\mathbf{x})| \le \frac{\|\phi\|_\rho}{\rho(1-\theta)}\|\mathbf{x}\|.$$

Next proposition will end the proof of Theorem 3.2.18

Proposition 3.2.20. *There exist* $\sigma>0$, $K>0$, $0<\varepsilon<1$ *such that for every* $\phi\in\mathcal{M}_\sigma$ *and for every* $n\in\mathbb{N}$, *we have*

$$\left\|T^n\phi(\mathbf{x})\right\| \le K(1-\varepsilon)^n\|\mathbf{x}\|,$$

for $\|\mathbf{x}\|<\sigma$.

Proof. Let us fix some constants $\widetilde{\Lambda}$, Λ, and $\varepsilon > 0$ such that

$$1 > \widetilde{\Lambda} > \Lambda > \rho(d\mathbf{f_0}) > 0 \quad \text{and } 1 - 3\varepsilon > \Lambda,$$

where ρ denotes the spectral radius.

For $\sigma > 0$ small enough, we have for every $\mathbf{x}$ such that $\|\mathbf{x}\| < \sigma$:

$$\begin{aligned} &|p(\mathbf{y})| \leq B\,\|\mathbf{y}\|\,, \\ &\|\mathbf{f}(\mathbf{x})\| \leq \Lambda\,\|\mathbf{x}\|\,, \\ &\Lambda\,(1 + \max\{|||\varepsilon|||_\sigma, |||\eta|||_\sigma\}) \leq 1 - 2\varepsilon, \end{aligned}$$

for a suitable $B > 0$, since ε, η and h are holomorphic and equal to 0 in $\mathbf{0}$. Moreover, B can be chosen as small as we want, up to a change of coordinates of the form $\mathbf{x} \mapsto (\mathbf{y}, bx_d)$.

Let us assume that

$$B \leq \min\{\Lambda, \varepsilon(1 - \widetilde{\Lambda})(\widetilde{\Lambda} - \Lambda)\}.$$

Now pick a $\phi \in \mathcal{M}_\sigma$, and let $A > 0$ such that $|\phi(\mathbf{x})| \leq A\,\|\mathbf{x}\|$ for $\|\mathbf{x}\|$ small enough, and hence $|||\phi|||_\sigma \leq A\sigma$ for σ small enough.

We want to estimate $\left|\frac{\partial\phi}{\partial x_d}(\mathbf{f}(\mathbf{x}))\right|$ for $\|\mathbf{x}\| < \sigma$.

Applying Lemma 3.2.19 (Cauchy's Lemma) twice, first to $\frac{\partial\phi}{\partial x_d}$, with $\theta = \Lambda/\widetilde{\Lambda}$ and $\rho = \sigma\widetilde{\Lambda}$ (and hence $\theta\rho = \Lambda\sigma$), and then to ϕ with $\theta = \widetilde{\Lambda}$ and $\rho = \sigma$, we get

$$\begin{aligned} \left|\frac{\partial\phi}{\partial x_d}(\mathbf{f}(\mathbf{x}))\right| &\leq \left|\frac{\partial\phi}{\partial x_d}(\mathbf{0})\right| + \frac{\left|\left|\left|\frac{\partial\phi}{\partial x_d}\right|\right|\right|_{\widetilde{\Lambda}\sigma}}{\widetilde{\Lambda}\sigma(1 - \Lambda\widetilde{\Lambda}^{-1})}\,\|\mathbf{f}(\mathbf{x})\| \\ &\leq \left|\frac{\partial\phi}{\partial x_d}(\mathbf{0})\right| + \frac{|||\phi|||_\sigma}{\sigma(1 - \widetilde{\Lambda})\widetilde{\Lambda}\sigma(1 - \Lambda\widetilde{\Lambda}^{-1})}\Lambda\,\|\mathbf{x}\| \\ &= \left|\frac{\partial\phi}{\partial x_d}(\mathbf{0})\right| + \frac{\Lambda|||\phi|||_\sigma}{\sigma^2(1 - \widetilde{\Lambda})(\widetilde{\Lambda} - \Lambda)}\,\|\mathbf{x}\|\,. \end{aligned}$$

Now we are ready to estimate $|T\phi(\mathbf{x})|$. For the first part of (3.26), we have, for $\|\mathbf{x}\| < \sigma$,

$$\left|\big(1 + \varepsilon(\mathbf{x})\big)\phi \circ \mathbf{f}(\mathbf{x})\right| \leq \big(1 + |||\varepsilon|||_\sigma\big)A\Lambda\,\|\mathbf{x}\|\,,$$

while for the integral part, we have

$$\begin{aligned} &|p(\mathbf{y})|\left|\int_0^1 \frac{\partial\phi}{\partial x_d}(\mathbf{f}(\mathbf{y}, tx_d))\big(1 + \eta(\mathbf{y}, tx_d)\big)dt\right| \\ &\quad\leq B\,\|\mathbf{y}\|\,\big(1 + |||\eta|||_\sigma\big)\left|\int_0^1 \frac{\partial\phi}{\partial x_d}(\mathbf{f}(\mathbf{y}, tx_d))dt\right| \\ &\quad\leq B\,\|\mathbf{y}\|\,\big(1 + |||\eta|||_\sigma\big)\left(\left|\frac{\partial\phi}{\partial x_d}(\mathbf{0})\right| + \frac{\Lambda|||\phi|||_\sigma}{\sigma^2(1 - \widetilde{\Lambda})(\widetilde{\Lambda} - \Lambda)}\,\|\mathbf{x}\|\right). \end{aligned}$$

So

$$
\begin{aligned}
|T\phi(\mathbf{x})| \leq & A \overbrace{\Lambda\big(1+\square\varepsilon\square_\sigma\big)}^{\leq 1-2\varepsilon} \|\mathbf{x}\| \\
& + B\,\|\mathbf{y}\|\,\big(1+\square\eta\square_\sigma\big)\left(\left|\frac{\partial\phi}{\partial x_d}(\mathbf{0})\right| + \frac{\Lambda\square\phi\square_\sigma\,\|\mathbf{x}\|}{\sigma^2(1-\widetilde{\Lambda})(\widetilde{\Lambda}-\Lambda)}\right) \\
\leq & A(1-2\varepsilon)\,\|\mathbf{x}\| + B\,\overbrace{\|\mathbf{y}\|}^{\leq\|\mathbf{x}\|}\,\big(1+\square\eta\square_\sigma\big)\left|\frac{\partial\phi}{\partial x_d}(\mathbf{0})\right| \\
& + \sigma^{-2}\,\overbrace{\|\mathbf{y}\|}^{<\sigma}\,\overbrace{\Lambda\big(1+\square\eta\square_\sigma\big)}^{\leq 1-2\varepsilon<1}\,\overbrace{\frac{B}{(1-\widetilde{\Lambda})(\widetilde{\Lambda}-\Lambda)}}^{\leq\varepsilon}\,\overbrace{\square\phi\square_\sigma}^{\leq A\sigma}\,\|\mathbf{x}\| \\
\leq & A(1-2\varepsilon)\,\|\mathbf{x}\| + B\big(1+\square\eta\square_\sigma\big)\left|\frac{\partial\phi}{\partial x_d}(\mathbf{0})\right|\,\|\mathbf{x}\| + A\varepsilon\,\|\mathbf{x}\| \\
\leq & \left(A(1-\varepsilon) + C\left|\frac{\partial\phi}{\partial x_d}(\mathbf{0})\right|\right)\|\mathbf{x}\|,
\end{aligned}
$$

where $C := B\big(1+\square\eta\square_\sigma\big) \leq 1-2\varepsilon$.

So we have proved that for every $\phi \in \mathcal{M}_\sigma$, if A is such that $|\phi(\mathbf{x})| \leq A\,\|\mathbf{x}\|$ when $\|\mathbf{x}\| < \sigma$, then there exists $C < 1$ such that

$$
|T\phi(\mathbf{x})| \leq \left(A(1-\varepsilon) + C\left|\frac{\partial\phi}{\partial x_d}(\mathbf{0})\right|\right)\|\mathbf{x}\| \tag{3.28}
$$

for $\|\mathbf{x}\| < \sigma$. We would like to apply (3.28) to $T\phi$ instead of ϕ, trying to get a recursion on the estimates. We need then to estimate $\frac{\partial T\phi}{\partial x_d}(\mathbf{0})$. We obtain

$$
\begin{aligned}
\frac{\partial T\phi}{\partial x_d}(\mathbf{0}) =& \frac{\partial\varepsilon}{\partial x_d}(\mathbf{0})\,\overbrace{\phi\circ\mathbf{f}(\mathbf{0})}^{=0} + \big(1+\overbrace{\varepsilon(\mathbf{0})}^{=0}\big)\frac{\partial\phi\circ\mathbf{f}}{\partial x_d}(\mathbf{0}) + \overbrace{p(\mathbf{0})}^{=0}\big(\ \ldots\ \big) \\
=& \frac{\partial\phi}{\partial x_d}(\mathbf{0})\frac{\partial f_d}{\partial x_d}(\mathbf{0}).
\end{aligned}
$$

Since

$$
\left|\frac{\partial f_d}{\partial x_d}(\mathbf{0})\right| \leq \rho(d\mathbf{f_0}) < \Lambda,
$$

we have

$$
\left|\frac{\partial T\phi}{\partial x_d}(\mathbf{0})\right| \leq \Lambda\frac{\partial\phi}{\partial x_d}(\mathbf{0}),
$$

and recursively

$$
\left|\frac{\partial T^n\phi}{\partial x_d}(\mathbf{0})\right| \leq \Lambda^n\frac{\partial\phi}{\partial x_d}(\mathbf{0}).
$$

Let us define recursively

$$A_0 := A, \quad A_{n+1} := (1-\varepsilon)A_n + C\Lambda^n \frac{\partial \phi}{\partial x_d}(\mathbf{0}); \tag{3.29}$$

applying recursively (3.28), we get that

$$\left| \frac{\partial T^n \phi}{\partial x_d}(\mathbf{x}) \right| \leq A_n \|\mathbf{x}\| .$$

Now we just want to estimate A_n using (3.29). But

$$\frac{A_{n+1}}{(1-\varepsilon)^{n+1}} = \frac{A_n}{(1-\varepsilon)^n} + \frac{C}{1-\varepsilon}\left(\frac{\Lambda}{1-\varepsilon}\right)^n \frac{\partial \phi}{\partial x_d}(\mathbf{0}).$$

It follows that

$$\begin{aligned} \frac{A_n}{(1-\varepsilon)^n} - A_0 &= \frac{C}{1-\varepsilon}\frac{\partial \phi}{\partial x_d}(\mathbf{0}) \sum_{k=0}^{n-1} \left(\frac{\Lambda}{1-\varepsilon}\right)^k \\ &\leq \frac{C}{1-\varepsilon}\frac{\partial \phi}{\partial x_d}(\mathbf{0}) \left(1 - \frac{\Lambda}{1-\varepsilon}\right)^{-1} . \end{aligned}$$

Hence

$$A_n \leq \left(A + \frac{C}{1-\varepsilon-\Lambda}\frac{\partial \phi}{\partial x_d}(\mathbf{0}) \right)(1-\varepsilon)^n = K(1-\varepsilon)^n,$$

where $K = A + \frac{C}{1-\varepsilon-\Lambda}\frac{\partial \phi}{\partial x_d}(\mathbf{0})$, and we are done. □

Proposition 3.2.20 is exactly the estimate we need to prove that ϕ defined by (3.27) is actually well defined and holomorphic, so it is an holomorphic solution of (3.24). Then using (3.25) we find $\widetilde{p}$, and we are done. □

Remark 3.2.21. Theorem 3.2.18 tells us that, given a $(d-1)$-reducible rigid germ $\mathbf{f} : (\mathbb{C}^d, 0) \to (\mathbb{C}^d, 0)$ with invertible internal action and only internal resonances, then we can change coordinates holomorphically in order to have that the last coordinate f_d of $\mathbf{f}$ is an affine function on x_d (with coefficients that depend on the other coordinates).

3.2.4. Remarks

Remark 3.2.22. To summarize what we have done in this section, let us consider q-reducible (dominant) attracting rigid germs $\mathbf{f} : (\mathbb{C}^d, 0) \to (\mathbb{C}^d, 0)$, with invertible internal action, of internal rank r and total rank s. Assume that the internal action $\underline{A}$ is of the form (3.2).

Then Theorem 3.2.4 (Poincaré-Dulac) gives directly the holomorphic classification when $q = 0$, *i.e.*, for invertible germs.

If we consider germs with $q = d$, then Theorem 3.2.15 gives the holomorphic classification (up to being sharper on resonant monomials in $\mathbf{g}$, see Remark 3.2.13, and on coefficients $\boldsymbol{\alpha}$), setting $r = s$ and $q = d$.

For germs with $q = d - 1$, Theorem 3.2.18 gives (more or less, see Remark 3.2.25) the classification in this case.

For germs with $1 \leq q \leq d - 2$, we can apply Theorem 3.2.6 and Theorem 3.2.15 to fix the first q coordinates; it remains to study what one can obtain for the last $d - q$ coordinates.

We shall summarize the results obtained for d and $(d-1)$-reducible germs in dimension d in the next two Corollaries.

Corollary 3.2.23. *Let $f : (\mathbb{C}^d, 0) \to (\mathbb{C}^d, 0)$ be an attracting d-reducible rigid germ with invertible internal action, of internal rank r. Up to taking iterates, and up to a holomorphic change of coordinates, we have*

$$\mathbf{f}(\mathbf{x}) = \Big(\boldsymbol{\lambda}\mathbf{x}_1, \boldsymbol{\alpha}\mathbf{x}^{\underline{B}}\big(1 + \mathbf{g}(\mathbf{x}_1)\big)\Big)^T, \tag{3.30}$$

where $\mathbf{x} = (\mathbf{x}_1, x_{r+1}, \ldots, x_d)$, $\boldsymbol{\lambda} \in (\mathbb{D}^*)^r$ is a vector made by the non-zero eigenvalues of $d\mathbf{f_0}$, $\boldsymbol{\alpha} \in (\mathbb{C}^*)^{d-r}$, $\mathbf{g}$ that contains only resonant monomials, and

$$\underline{B} = \big(\ \underline{C} \mid \underline{D}\ \big) \in \mathcal{M}(d - r \times d, \mathbb{N})$$

is such that $\underline{D}$ is invertible, and every row $\mathbf{d}_i$ of $\mathcal{D}$ is such that $|\mathbf{d}_i| = \sum_j \mathbf{d}_i^j \geq 1$ and $\mathbf{d}_i \neq \mathbf{e}_i$.

Finally, we can suppose $\|\boldsymbol{\alpha}\|$ arbitrarily small, while the coordinates of $\boldsymbol{\alpha}$ have to satisfy $m_g(1)$ conditions, where $m_g(1)$ is the geometric multiplicity of 1 for $\underline{D}$.

Proof. The statement directly follows from Theorem 3.2.15 applied to $r = d$. For the conditions on $\underline{D}$, they come from directly checking the critical set of $\mathbf{f}$ as in (3.30), while the computation on α follows from studying these germs up to a linear change of coordinates. □

Corollary 3.2.24. *Let $f : (\mathbb{C}^d, 0) \to (\mathbb{C}^d, 0)$ be an attracting $(d-1)$-reducible rigid germ with invertible internal action, of internal rank r and total rank s. Up to taking an iterate, and up to a holomorphic change of coordinates, we have*

$$\mathbf{f}(\mathbf{x}) = \Big(\boldsymbol{\lambda}\mathbf{x}_1, \boldsymbol{\alpha}\mathbf{y}^{\underline{B}}\big(1 + \mathbf{g}(\mathbf{x}_1)\big), \mu\mathbf{y}^{\mathbf{l}}x_d + p(\mathbf{y})\Big)^T, \tag{3.31}$$

with $\mathbf{x} = (\mathbf{y}, x_d)^T$, $\mathbf{y} = (\mathbf{x}_1, x_{r+1}, \ldots, x_{d-1})^T$, $\mathbf{l} \in \mathcal{M}(1 \times d-1, \mathbb{N})$, $\mathbf{g}$ *with only internal resonant monomials, and* $p(\mathbf{y}) \in \mathfrak{m}$ *is a suitable analytic function.*

If moreover $r = s$, *and hence* $\mathbf{l} = \mathbf{0}$, *we have that* $p(\mathbf{y})$ *is a polynomial with only resonant monomials (for the last coordinate).*

Proof. It follows directly from Theorem 3.2.18. □

Remark 3.2.25. While studying $(d-1)$-reducible rigid germs in dimension d (with only internal resonances), following the proof of Theorem 2.4.3 as given in [25], we should consider change of coordinates of the form

$$\Phi(\mathbf{x}) = (\mathbf{y}, x_d + q(\mathbf{y}))^T \tag{3.32}$$

(we are using the notations of Corollary 3.2.24). In dimension 2, one can obtain (holomorphically) that p is a polynomial in $\mathbf{y} = x_1$. This is no longer true, not even formally, in higher dimensions. Indeed, by computing the coefficients in the conjugacy relation, on can show that there are infinitely many coefficients of p that cannot be taken as 0 up to a change of coordinates of the form (3.32). It can be also shown that, in order to maintain the normal form as in (3.31), one can (basically) consider only change of coordinates such as (3.32).

We could also have been sharper for example on checking which conditions one can impose on $\boldsymbol{\alpha}$ on (3.31): actually one can achieve the same results as in the case of Corollary 3.2.23, with the same techniques, but since the classification would not be complete as well, we did not put it in the statement of Corollary 3.2.24.

3.3. Rigid germs in dimension 3

In this section with the next theorem, we shall summarize what we have proved for (dominant) attracting rigid germs in dimension 3. We shall omit the transpositions on vectors.

Theorem 3.3.1. *Let* $f : (\mathbb{C}^3, 0) \to (\mathbb{C}^3, 0)$ *be an attracting (dominant)* q*-reducible rigid germ with invertible internal action, of internal rank* r *and total rank* s. *Let us denote by* $\lambda_1, \ldots, \lambda_s$ *the non-zero eigenvalues of* df_0.

Then up to holomorphic change of coordinates (and up to replacing f *by* f^2 *when* $r = 2$*), we have:*

$q = 0$: *we order* λ_i *such that* $|\lambda_1| \geq |\lambda_2| \geq |\lambda_3|$. *Then*

$$\mathbf{f}(\mathbf{x}) = \big(\lambda_1 x_1, \lambda_2 x_2 + g_2(x_1), \lambda_3 x_3 + g_3(x_1, x_2)\big).$$

If $\lambda_2 = \lambda_1^n$ *for a suitable* $n \in \mathbb{N}^*$, *then* $g_2(x_1) = \varepsilon x_1^n$; *moreover, if* $\lambda_3 = \lambda_1^m$ *for a suitable* $m \in \mathbb{N}^*$, *then* $g_3(x_1, x_2)$ *is a homogeneous polynomial of degree* m, *where* x_1 *has weight* 1 *and* x_2 *has weight* n, *while if* $\lambda_3 \neq \lambda_1^m$ *for every* $m \in \mathbb{N}^*$, *then* $g_3 \equiv 0$.
If $\lambda_2 \neq \lambda_1^n$ *for every* $n \in \mathbb{N}^*$, *then* $g_2 \equiv 0$; *moreover, if* $\lambda_3 = \lambda_1^n \lambda_2^m$ *for suitable* $n, m \in \mathbb{N}$ *(not both* 0*), then* $g_3(x_1, x_2) = \varepsilon x_1^n x_2^m$, *otherwise* $g_3 \equiv 0$.
Finally, if $\lambda_1 = \lambda_2$ *or* $\lambda_2 = \lambda_3$, *we can suppose that the coefficients of* g_2 *and* g_3 *are such that* df_0 *is in Jordan form.*

$q = 1$: *we split the classification with respect to the values of* r *and* s.

$r = 1$: *if* $s = 2$ *then up to holomorphic conjugacy we have*

$$\mathbf{f}(\mathbf{x}) = \big(\lambda x_1, \mu x_2 + \delta x_1^u, \gamma x_1^l x_3 + p(x_1, x_2)\big),$$

where λ, μ *are the two non-zero eigenvalues of* df_0, $\delta \in \mathbb{C}$ *if* $\lambda^u = \mu$, *while* $\delta = 0$ *otherwise; moreover* $l \geq 1$.
If $s = 1$ *then we have*

$$\mathbf{f}(\mathbf{x}) = \big(\lambda_1 x_1, g(\mathbf{x}), h(\mathbf{x})\big)$$

for suitable g, h.

$r = 0$: *if* $s = 2$ *then up to holomorphic conjugacy we have*

$$\mathbf{f}(\mathbf{x}) = \big(x_1^a, \lambda x_2, \mu x_3 + \delta x_2^u\big),$$

where λ, μ *are the two non-zero eigenvalues of* df_0, $\delta \in \mathbb{C}$ *if* $\lambda^u = \mu$, *while* $\delta = 0$ *otherwise; moreover* $a \geq 2$.
If $s = 1$ *then up to holomorphic conjugacy we have*

$$\mathbf{f}(\mathbf{x}) = \big(x_1^a, \lambda x_2, \gamma x_1^l x_3 + p(x_1, x_2)\big),$$

where λ *is the non-null eigenvalue of* df_0, $a \geq 2$ *and* $l \geq 1$.
If $s = 0$ *then we have*

$$\mathbf{f}(\mathbf{x}) = \big(x_1^a, g(\mathbf{x}), h(\mathbf{x})\big)$$

for suitable g, h, *and* $a \geq 2$.

$q = 2$: *we split the classification with respect to the values of* r *and* s.

$r = 2$: *then* $s = 2$ *and*

$$\mathbf{f}(\mathbf{x}) = \big(\lambda_1 x_1, \lambda_2 x_2, x_1^l x_2^m x_3 + p(x_1, x_2)\big),$$

where $l, m \geq 1$ *and* p *is a suitable holomorphic map in* $\mathfrak{m}^2$.

$r = 1$: *if* $s = 2$ *we have*

$$\mathbf{f}(\mathbf{x}) = \big(\lambda x_1, \mu x_2 + \delta x_1^u, x_1^a x_3^c\big),$$

where $a \geq 1$, $c \geq 2$, λ, μ *are the two non-zero eigenvalues of* df_0, $\delta \in \mathbb{C}$ *if* $\lambda^u = \mu$, *while* $\delta = 0$ *otherwise.*
If $s = 1$ *then we have*

$$\mathbf{f}(\mathbf{x}) = \big(\lambda x_1, x_1^a x_2^b, x_1^l x_2^m x_3 + p(x_1, x_2)\big),$$

where $l + m \geq 1$, $a + l \geq 1$, $b \geq 1$, $b + m \geq 2$, *and* $p(x_1, x_2) - \nu x_2 \in \mathfrak{m}^2$ *for a suitable* ν*).*

$r = 0$: *if* $s = 1$, *then*

$$\mathbf{f}(\mathbf{x}) = \Big(\alpha_1 x_1^a x_2^b (1 + \eta x_3^v), \alpha_2 x_1^c x_2^d, \lambda x_3\Big), \tag{3.33}$$

where $ad \neq bc$, $c + d \geq 2$, $\max\{a - 1, b\} \geq 1$, *and if* v *satisfies the resonance relation* $(\mu^v - a)(\mu^v - d) = bc$, $\eta \in \mathbb{C}$ *(*$\eta = 0$ *otherwise).*
If $s = 0$, *then*

$$\mathbf{f}(\mathbf{x}) = \Big(\alpha_1 x_1^a x_2^b, \alpha_2 x_1^c x_2^d, \gamma x_1^l x_2^m x_3 + p(x_1, x_2)\Big),$$

where $ad \neq bc$, $c + d \geq 2$, $\max\{a - 1, b\} \geq 1$, $l + m \geq 1$ *and* $p \in \mathfrak{m}^2$ *is a suitable holomorphic map.*

$q = 3$: *we split the classification with respect to the values of* $r = s$.

$r = 2$: *then*

$$\mathbf{f}(\mathbf{x}) = \big(\lambda_1 x_1, \lambda_2 x_2, x_1^a x_2^b x_3^c\big),$$

with $a, b \geq 1$, $c \geq 2$.

$r = 1$: *then*

$$\mathbf{f}(\mathbf{x}) = \big(\lambda_1 x_1, \alpha_1 x_1^{a_1} x_2^{b_1} x_3^{c_1} (1 + \eta x_1^n), \alpha_2 x_1^{a_2} x_2^{b_2} x_3^{c_2}\big),$$

with $a_1, a_2 \in \mathbb{N}$, $b_1 c_2 \neq b_2 c_1$, $a_1 + a_2 \geq 1$, $b_1 + b_2 \geq 2$, $c_1 + c_2 \geq 2$, *and if* $(\lambda_1^n - b_1)(\lambda_1^n - c_2) = b_2 c_1$, *then* $\eta \in \mathbb{C}$, *otherwise* $\eta = 0$.

$r = 0$: *then*

$$\mathbf{f}(\mathbf{x}) = \boldsymbol{\alpha}\mathbf{x}^{\underline{A}},$$

with $\boldsymbol{\alpha} \in (\mathbb{C}^*)^3$, *and* $\underline{A} = (a_i^j) \in \mathcal{M}(3 \times 3, \mathbb{N})$ *is the internal action (or equivalently the principal part) of* f, *such that* $a_i^1 + a_i^2 + a_i^3 \geq 2$ *for every* $i = 1, 2, 3$.

Proof. The proof is straightforward using Theorem 3.2.4 (Poincaré-Dulac), Theorem 3.2.15, and Theorem 3.2.18 for $q = 2$. Here the resonance relations have been made explicit in each case (see also Remark 3.2.13). Some considerations on coefficients arise by changes of coordinates of the form $\mathbf{x} \mapsto \gamma\mathbf{x}$. □

Chapter 4
Construction of non-Kahler 3-folds

4.1. Kato surfaces, rigid germs and Hénon maps

4.1.1. Kato surfaces

Masahide Kato gives (see [40]) a method to construct compact complex surfaces of class VII_0 with $b_2 > 0$ that admit a global spherical shell, the so called *Kato surfaces*.

Definition 4.1.1. Let $B = \overline{\mathcal{B}_\varepsilon}$ be a closed ball in $\mathbb{C}^2$ of center 0 and radius $\varepsilon > 0$, and $\pi : \widetilde{B} \to B$ a modification over 0. Let $\sigma : B \to \widetilde{B}$ be a biholomorphism with its image such that $\sigma(0)$ is a point of the exceptional divisor of π. The couple (π, σ) is called a **Kato data**.

Remark 4.1.2. From a Kato data (π, σ) we can construct a manifold X. If $B = \mathcal{B}_\varepsilon$ is a ball in $\mathbb{C}^2$ of center 0 and radius $\varepsilon > 0$, roughly speaking we take $\widetilde{X} = \pi^{-1}(\overline{B}) \backslash \sigma(B)$, and take the quotient by $\sigma \circ \pi : \pi^{-1}(\partial B) \to \sigma(\partial B)$, that is a biholomorphism (for ε small enough). This is the main idea, for the precise construction (in dimension 3) see Subsection 4.2.3. The manifold so constructed admits evidently a global spherical shell; namely, it is the natural embedding of a neighborhood of $\pi^{-1}(\partial B)$ (or equivalently of $\sigma(\partial B)$).

Definition 4.1.3. Let (π, σ) be a Kato data. The compact complex surface constructed as in 4.1.2 is the **Kato surface** associated to the given Kato data.

Definition 4.1.4. Let (π, σ) be a Kato data. Then we can consider $f_0 = \pi \circ \sigma : B \to B$, that turns out to be a holomorphic germ, with a fixed point in 0 the center of B. We shall call this germ the **base germ** associated to the given Kato data.

In general, given a holomorphic germ $f_0 : (\mathbb{C}^n, 0) \to (\mathbb{C}^n, 0)$, we shall call **resolution** for f_0 a decomposition $f_0 = \pi \circ \sigma$, with π a modification over 0 and σ a (germ) biholomorphism that sends 0 into a point of the exceptional divisor of π. We shall also denote $f_0 = (\pi, \sigma)$ if we want to stress the Kato data instead of the base germ.

Dloussky and others studied this phenomenon very deeply. Here we just want to list a few results on Kato surfaces (see [15] for proofs).

Theorem 4.1.5. *Let $f_0 = (\pi, \sigma)$ a Kato data and X the Kato surface associated to it. Then the following statements hold.*

(i) *X is a Class VII_0 surface.*
(ii) *X admits a global spherical shell.*
(iii) $b_2(X) = \mathrm{weight}(\pi)$.
(iv) *X has exactly $b_2(X)$ rational curves. It has also an elliptic curve if and only if f_0 is holomorphically conjugated to $(x, y) \mapsto (\alpha x, x^c y)$. There are no other (compact) curves.*

4.1.2. Hénon maps

Polynomial automorphisms of $\mathbb{C}^2$ can be subdivided into two classes, *elementary automorphisms*, whose dynamics is easier to study, and compositions of Hénon maps (see [29]).

Definition 4.1.6. An automorphism $f : \mathbb{C}^2 \to \mathbb{C}^2$ is said to be a **Hénon map** if it is of the form

$$f(x, y) = \big(p(x) - ay, x\big), \tag{4.1}$$

with p a polynomial of degree $d = \deg p \geq 2$.

Remark 4.1.7. Consider a Hénon map such as in (4.1), with $p(x) = p_0 + \dots p_d x^d$. Up to an affine change of coordinates, a Hénon map can be chosen such that p is monic (*i.e.*, $p_d = 1$), and $p_{d-1} = 0$.

Remark 4.1.8. Let us consider a Hénon map $f : \mathbb{C}^2 \to \mathbb{C}^2$ as in (4.1), and its extension $F : \mathbb{P}^2 \to \mathbb{P}^2$ to $\mathbb{P}^2$. Then, computing the indeterminacy sets I^+ and I^- of F and F^{-1} respectively, one gets

$$\begin{aligned} I^+ &= \{[0 : 1 : 0]\}, \\ I^- &= \{[1 : 0 : 0]\}. \end{aligned}$$

In particular, $I^+ \cap I^- = \emptyset$.

In order to extend some results for Hénon maps in higher dimensions, one need some candidates to replace Hénon maps, since the structure of polynomial automorphisms of $\mathbb{C}^n$ is not yet completely understood.

One property of Hénon maps has been identified, by Sibony (see [57]) and many others, in order to extend several results to higher dimensions: regularity.

Definition 4.1.9. Let $f : \mathbb{C}^n \to \mathbb{C}^n$ be a polynomial automorphism, $F : \mathbb{P}^n \to \mathbb{P}^n$ be its extension to $\mathbb{P}^n$, and denote by I^+ and I^- the indeterminacy sets of F and F^{-1} respectively. Then f is said to be **regular** (in the sense of Sibony) if $I^+ \cap I^- = \emptyset$.

Dloussky and Oeljeklaus (see [17]) studied the case when the germ f_0 arises from the action at infinity of an automorphism of $\mathbb{C}^2$. Indeed, starting from a Hénon map f, taking the action F induced in $\mathbb{P}^2$ and a fixed point p in the line at infinity, one can find a germ $f_0 := F_p$, and a resolution $f_0 = \pi \circ \sigma$ to construct a Kato surface.

We shall see in detail an example of this phenomenon.

Example 4.1.10. Let us take as example, for the sake of simplicity, the simplest Hénon map:

$$f(x, y) = (x^2 - y, x).$$

Then its induced map in $\mathbb{P}^2$ is $F : \mathbb{P}^2 \to \mathbb{P}^2$ given by

$$F[x : y : t] = [x^2 - yt : xt : t^2].$$

Computing the inverse of f, we get

$$f^{-1}(x, y) = (y, y^2 - x)$$

and

$$F^{-1}[x : y : t] = [yt : y^2 - xt : t^2].$$

Studying the action of F at infinity, we see that the indeterminacy set for F is $I^+ = \{[0 : 1 : 0]\}$, while the indeterminacy set of F^{-1} is $I^- = \{[1 : 0 : 0]\}$. Moreover, denoting by p the point $[1 : 0 : 0]$, we have that the dynamics of F in $\{t = 0\}$ is very simple: $F(q) = p$ for every $q \notin I^+$, while $F[0 : 1 : 0]$ is the whole line at infinity. In particular p is a fixed point for F; let us consider the germ $f_0 := F_p$. Choosing the chart $\{x \neq 0\}$, we get

$$f_0(y, t) = \left(\frac{t}{1 - yt}, \frac{t^2}{1 - yt}\right).$$

We want now to find a resolution for f_0.

The strict germ f_0 fails to be a biholomorphism in its critical set $\{t = 0\}$: outside it is clearly a biholomorphism with its image, since it is given by an automorphism of $\mathbb{C}^2$. So the idea is to blow-up the image through f_0 of the line $E_0^{(0)} := \{t = 0\}$. Since in this case $f_0(y, 0) = 0$, we blow-up the origin. So let us consider the blow-up π_1 in 0, and we choose coordinates (y_1, t_1) such that $(y, t) =: (y_0, t_0) = \pi_1(y_1, t_1) = (y_1, y_1 t_1)$. Hence we get

$$f_1(y,t) := \pi_1^{-1} \circ f_0(y,t) = \left(\frac{t}{1-yt}, t\right).$$

In this case we have two divisors, $E_0^{(1)} = \{t_1 = 0\}$ and $E_1^{(1)} = \{y_1 = 0\}$ in these coordinates. Here $f_1(\{t = 0\}) = 0$ the intersection of these two divisors, so we blow-up it, and call the blow-up π_2. We choose coordinates (y_2, t_2) such that $(y_1, t_1) = \pi_2(y_2, t_2) = (y_2 t_2, t_2)$, and we get

$$f_2(y,t) := \pi_2^{-1} \circ f_1(y,t) = \left(\frac{1}{1-yt}, t\right).$$

Another exceptional component $E_2^{(2)} = \{t_2 = 0\}$ arises, while $E_i^{(1)}$ lifts to their strict transforms $E_i^{(2)}$ (for $i = 0, 1$). In this case $f_2(\{t = 0\}) = (1, 0)$, that is a free point in $E_2^{(2)}$. We first change coordinates by a translation $(y_2, t_2) = \widetilde{\tau}_2(\widetilde{y}_2, \widetilde{t}_2) = (\widetilde{y} + 1, \widetilde{t})$, and we get

$$\widetilde{f}_2 : (y,t) = \widetilde{\tau}_2^{-1} \circ f_2(y,t) = \left(\frac{yt}{1-yt}, t\right),$$

and then we blow-up the origin in these coordinates, by taking $(\widetilde{y}_2, \widetilde{t}_2) = \widetilde{\pi}_3(y_3, t_3) = (y_3 t_3, t_3)$. Let us denote $\pi_3 = \widetilde{\tau}_2 \circ \widetilde{\pi}_3$; then we get

$$f_3(y,t) := \pi_3^{-1} \circ f_2(y,t) = \left(\frac{y}{1-yt}, t\right).$$

Here we have a new exceptional divisor $E_3^{(3)} = \{t_3 = 0\}$, and as before the strict transforms of the former three divisors, that we shall denote by $E_i^{(3)}$ for $i = 0, 1, 2$. Moreover f_3 is a (germ) biholomorphism that takes $(y, t) = 0$ into the smooth point $(y_3, t_3) = 0$, that belongs only to $E_3^{(3)}$.

Hence, defining

$$\pi = \pi_1 \circ \pi_2 \circ \pi_3 \qquad \text{and } \sigma = f_3$$

we have found the resolution of $f_0 = \pi \circ \sigma$.

We can construct now the Kato surface associated to $f_0 = \pi \circ \sigma$: by the identification of the construction, two divisors, $E_0^{(3)}$ and $E_3^{(3)}$, glue together giving a divisor E_0, while the other divisors $E_i^{(3)}$ for $i = 1, 2$ are not affected, and each one gives us a divisor that we shall call E_i.

We shall denote by E the union of E_i for $i = 0, \ldots, 2$ (thought as curves in X).

4.1.3. Dynamical properties of some Kato surfaces

Remark 4.1.11. When we get a Kato data from a polynomial automorphism of $\mathbb{C}^2$ as we have seen in Example 4.1.10, the Kato surface associated has a dynamical interpretation. Indeed, we can consider the basin of attraction $U_f(p)$ of $f : \mathbb{C}^2 \to \mathbb{C}^2$ in p, that is an open subset of $\mathbb{C}^2$. This open set contains a neighborhood B of p (without the line at infinity), being f_0 an attracting germ.

Since π is a biholomorphism outside the exceptional divisor, and $\sigma^{-1}(E_3^{(3)} \cap \sigma(B)) = \{t = 0\} \cap B$, to identify through the action of f_0 means, in $X \setminus E$, to identify through f, and hence to take a fundamental domain for $U_f(p)$. So the Kato surface X can be seen as a compactification of a fundamental domain for $U_f(p)$, by adding E (that in this case is made by 3 irreducible rational curves).

Question 4.1.12. At this point, two natural questions arise.

(i) Does there exist the compactification to a point Y of the fundamental domain $X \setminus E$ such that Y is a (singular) surface?
(ii) Can we obtain some informations on X by studying the local dynamics of the germ f_0?

The first question is equivalent to asking if we can contract the divisor E to a point, thus obtaining a compact complex surface (possibly singular in the point where we contracted). Let us define properly this phenomenon.

Definition 4.1.13. Let X be a n-manifold, and $E_X \hookrightarrow X$ (the support of) a divisor. Consider then E_Y a k-variety (resp., $E_Y = \{p\}$), and the diagram

$$\begin{array}{ccc} X & & \\ \uparrow & & \\ E & \longrightarrow & E_Y \end{array}$$

A **contraction** (resp., to a point) is a (possibly singular) n-variety Y and arrows such that the diagram

$$\begin{array}{ccc} X & \longrightarrow & Y \\ \uparrow & & \uparrow \\ E_X & \longrightarrow & E_Y \end{array}$$

commutes.

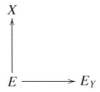

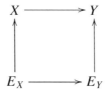

Then there is a result by Grauert that gives us the answer in the 2-dimensional case.

Theorem 4.1.14 ([32, Chapter 4, Section 8e), pp. 366–367]). *Let X be a surface and $E \subset X$ (the support of) a divisor. If the intersection matrix of E is negative defined, then there exists the contraction of E to a point.*

Thanks to this result and direct computation, it can be seen that for Kato surfaces, one can always contract the exceptional divisor.

Theorem 4.1.15 ([15, Part II, 2]). *Let X be the Kato surface associated to a Kato data $f = (\pi, \sigma)$, and E the exceptional divisor of π plus the critical set for f. Then there exists the contraction of E to a point.*

Let us show the computation of the intersection matrix of E of Example 4.1.10. We first need a lemma.

Lemma 4.1.16. *Let X be a surface, $p \in X$ a point, and $\pi : \widetilde{X} \to X$ the blow-up of X along p, with exceptional divisor E. Then we have*

(i) $E \cdot E = -1$,
(ii) *if p is a smooth point of an irreducible curve D, we have that $\widetilde{D} \cdot \widetilde{D} = D \cdot D - 1$, where $\widetilde{D}$ is the strict transform of D through π.*

Proof. Thanks to Proposition 1.2.25, we have that

$$0 = \pi^*(D) \cdot E = \widetilde{D} \cdot E + E \cdot E,$$

and since $\widetilde{D}$ and E meet transversely, we have $\widetilde{D} \cdot E = 1$ and hence $E \cdot E = -1$.

Thanks to Proposition 1.2.24, we have that

$$D \cdot D = \pi^* D \cdot \pi^* D = \widetilde{D} \cdot \widetilde{D} + 2\widetilde{D} \cdot E + E \cdot E = \widetilde{D} \cdot \widetilde{D} - 1. \qquad \square$$

Example 4.1.17. Let us make the computation of the intersection matrix of E in the case of Example 4.1.10. We proceed step by step by blow-up.

- We first start with $E_0^{(0)}$ the line (at infinity) in $\mathbb{P}^2$, that has self-intersection $E_0^{(0)} \cdot E_0^{(0)} = 1$.
- Then we blow-up a smooth point in $E_0^{(0)}$, and thanks to Lemma 4.1.16 we get $E_1^{(1)} \cdot E_1^{(1)} = -1$ and $E_0^{(1)} \cdot E_0^{(1)} = 0$.
- Now we blow-up the point in the intersection of the two divisors, and hence $E_2^{(2)} \cdot E_2^{(2)} = -1$, $E_1^{(2)} \cdot E_1^{(2)} = -2$ and $E_0^{(2)} \cdot E_0^{(2)} = -1$.

- Finally we blow-up a smooth point in $E_2^{(2)}$, getting $E_3^{(3)} \cdot E_3^{(3)} = 1$, $E_2^{(3)} \cdot E_2^{(3)} = -2$, while the last two do not change, and hence $E_1^{(3)} \cdot E_1^{(3)} = -2$ and $E_0^{(3)} \cdot E_0^{(3)} = -1$.

To understand what happens when we glue, first of all the identification does not affect $E_i^{(3)}$ for $i = 1, 2$, and hence $E_1 \cdot E_1 = -2$ and $E_2 \cdot E_2 = -2$. To compute $E_0 \cdot E_0$, we have to look at the situation in the universal covering of X. Indeed we obtain that E_0 is the strict transform of $E_3^{(3)}$ by two blow-ups, and hence $E_0 \cdot E_0 = E_3^{(3)} \cdot E_3^{(3)} - 2 = -3$. In general, if we have n blow-ups, and hence a gluing between $E_0^{(0)}$ and $E_n^{(n)}$, we get $E_0 \cdots E_0 = (E_n^{(n)})^2 + \left(E_0^{(n)}\right)^2 - \left(E_0^{(0)}\right)^2$. See Subsection 4.2.3 for the construction of the universal covering, and Subsection 4.3.3(Gluing) for the analogous situation in dimension 3.

Hence we get the matrix

$$\begin{pmatrix} -3 & 0 & 2 \\ 0 & -2 & 1 \\ 2 & 1 & -2 \end{pmatrix},$$

that is negative defined.

We shall focus now on Question 4.1.12.(ii). One can easily see is that if $f_0 = (\pi, \sigma)$ is a Kato data, then f is a (irreducible) strict germ. Then there is a sort of "dictionary" between dynamical objects for $f : (\mathbb{C}^2, 0) \to (\mathbb{C}^2, 0)$ and geometrical objects for its Kato surface X. See [15, Part II, Chapter 0] and [16] for further details.

Proposition 4.1.18. *Let X be the Kato surface associated to a strict germ $f = \pi \circ \sigma : (\mathbb{C}^2, 0) \to (\mathbb{C}^2, 0)$. Then*

(i) *each invariant curve C for f_0 corresponds to a (elliptic) curve on X;*
(ii) *invariant vector fields for f_0 correspond to global vector fields on X;*
(iii) *invariant foliations for f_0 correspond to global (possibly singular) foliations on X.*

Starting from a Hénon map f, we get an explicit formula for f_0, and Favre showed in [25] that f_0 is an attracting strict germ of class 4, and gave the normal forms for such germs (see Theorem 2.4.3). In particular we get:

Theorem 4.1.19 ([25, Proposition 2.2]). *Let $f : \mathbb{C}^2 \to \mathbb{C}^2$ a Hénon map, of the form*

$$f(x, y) = \big(p(x) - ay, x\big), \tag{4.1}$$

where $\deg p = d$ *and* $a \in \mathbb{C}^*$. *Consider f_0 the germ in the fixed point at infinity* $[1 : 0 : 0]$ *for the extension $F : \mathbb{P}^2 \to \mathbb{P}^2$ of f. Then f_0 is holomorphically conjugated to*

$$\widetilde{f}(x, y) = \left(x^d, \frac{a}{d}x^{2d-2}y + x^{d-1}\big(1 + R(x)\big)\right). \tag{4.2}$$

Remark 4.1.20. Thanks to Theorem 4.1.19, we get two things. First, the normal form (4.1.19) has the first coordinate that depends only on x, and hence there exists a foliation, namely the one given by dx, that is invariant for f. One can see directly from the proof of Theorem 4.1.19 that $\{x = 0\}$ is the line at infinity in $\mathbb{P}^2$. This foliation induces a foliation $\mathcal{F}$ on the Kato surface X which has E as a leaf (see 4.1.18).

Moreover, the normal form (4.2) has another property: for every x fixed, the second coordinate is affine on y: this gives a special way how to go from a leaf to another, and allows also to compute the first fundamental group of the basin of attraction, as we shall see in the next theorem.

Theorem 4.1.21 ([37, Sections 7, 8], see also [25, Theorem 2.4]). *Let $f\colon \mathbb{C}^2 \to \mathbb{C}^2$ be a Hénon map*

$$f(x, y) = \big(p(x) - ay, x\big), \tag{4.1}$$

where $\deg p = d$ *and* $a \in \mathbb{C}^*$. *Let us denote by Ω the basin of attraction for f at $p = [1 : 0 : 0]$. Then*

$$\pi_1(\Omega) \cong \mathbb{Z}\left[\frac{1}{d}\right] := \left\{\frac{m}{d^n} : m \in \mathbb{Z} \text{ and } n \in \mathbb{N}\right\}.$$

4.2. Construction in the 3D case

4.2.1. The example

In the following we are going to analyze the construction of a Kato 3-fold associated to the action at infinity at a suitable point p for a polynomial automorphism in $\mathbb{C}^3$. Let us consider

$$f(x, y, z) = \big(x^2 + cy^2 + z, y^2 + x, y\big), \tag{4.3}$$

with $c \in \mathbb{C}$. Its inverse is

$$f^{-1}(x, y, z) = \big(y - z^2, z, x - (y - z^2)^2 - cz^2\big).$$

Hence f is a polynomial automorphism of degree 2, with a polynomial inverse f^{-1} of degree 4.

We extend f and f^{-1} to $\mathbb{P}^3 = \{[x : y : z : t]\}$; then we get

$$\begin{aligned} F[x : y : z : t] &= [x^2 + cy^2 + zt : y^2 + xt : yt : t^2], \\ F^{-1}[x : y : z : t] &= \left[yt^3 - z^2t^2 : zt^3 : xt^3 - (yt - z^2)^2 - cz^2t^2 : t^4\right]. \end{aligned}$$

The action of F and F^{-1} on the plane at infinity $\{t = 0\}$ is given by

$$\begin{aligned} F[x : y : z : 0] &= [x^2 + cy^2 : y^2 : 0 : 0], \\ F^{-1}[x : y : z : 0] &= [0 : 0 : -z^4 : 0]; \end{aligned}$$

in particular the sets of indeterminacy of F and F^{-1} are given by

$$\begin{aligned} I^+ &= \{t = x = y = 0\} = \{[0 : 0 : 1 : 0]\}, \\ I^- &= \{t = z = 0\}. \end{aligned}$$

Remark 4.2.1. We clearly have $I^+ \cap I^- = \emptyset$, *i.e.* F and F^{-1} are regular (in the sense of Sibony).

We want to focus our attention on F^{-1}, and its behavior at infinity. The action of $F^{-1}|_{t=0}$ is very simple, everything (besides the line I^-) is contracted to the only fixed point $p = [0 : 0 : 1 : 0]$ (that is also the only point in I^+).

Let us consider now the germ $f_0 = F_p^{-1}$; considering the chart of $\mathbb{P}^3$ given by $\{z \neq 0\}$, after conjugating by $(x, y, t) \mapsto (x, y, -t)$, we have

$$f_0(x, y, t) = \frac{\left(t^2(1 + yt), t^3, t^4\right)}{(1 + yt)^2 + ct^2 + xt^3}. \tag{4.4}$$

Remark 4.2.2. The structure of polynomial automorphisms in $\mathbb{C}^d$ for $d \geq 3$ is not well known (opposite to the 2-dimensional case, with the factorization into Hénon maps and elementary maps, see Subsection 4.1.2), and only a few cases have been classified. For instance, the automorphisms of degree 2 in $\mathbb{C}^3$ has been classified up to affine conjugacy: see [30] and [49].

Moreover by direct computation, one can see that regular automorphisms belong to classes

$$\begin{aligned} H_4 : f(x, y, z) &= \big(P(x, y) + az, Q(y) + x, y\big), \\ H_5 : f(x, y, z) &= \big(P(x, y) + az, Q(x) + by, x\big), \end{aligned}$$

with $a, b \neq 0$, $\deg P = \deg Q = 2$ and $P(x, 0) \neq 0$ (see [30, Theorem 2.1]).

With respect to the classification given in [49], regular automorphisms belong to class H_1 and H_2 of shift-like mappings (2-shift), of the form

$$f(x, y, z) = \big(y, z, y^2 + L(y, z) + dx\big) \circ \big(y, z, P'(y, z) + d'x\big), \quad (4.5)$$

with $d, d' \neq 0$, and suitable polynomials L, P' with $\deg L = 1$, $\deg P' = 2$.

In particular, our example (4.3) belongs to class H_4, with $a = 1$, $P(x, y) = x^2 + cy^2$ and $Q(y) = y^2$, and up to permuting coordinates can be decomposed as in (4.5). Indeed, if we denote by f the polynomial automorphism as in (4.3) and by σ the permutation $\sigma(x, y, z) = (z, x, y)$, we have

$$\sigma \circ f \circ \sigma^{-1}(x, y, z) = \big(y, z, y^2 + x\big) \circ \big(y, z, y^2 + cz^2 + x\big).$$

Remark 4.2.3. The polynomial automorphism (4.3) has been considered also by Oeljeklaus and Renaud (see [52]). They studied the basin of attraction by f to the line at infinity I^-, and constructed a Class L 3-variety (introduced by Kato, see [41]).

Here we are considering the inverse map f^{-1} to construct a Kato 3-fold, and to study its algebraic and dynamical properties in detail.

4.2.2. Resolution of f_0

In order to get a Kato data from f_0, our goal now is to find a resolution of f_0, *i.e.*, to decompose f_0 as a composition of two maps, $\pi \circ \sigma$, where π is a modification over p, and σ is an automorphism between a neighborhood of p and a neighborhood of a suitable point $q \in \pi^{-1}(p)$.

Our strategy is then to blow-up the image of $E_0^{(0)} := \{t = 0\} = \mathcal{C}(f_0)$ with respect to f_0, to consider the lifted map $f_1 = \pi_1^{-1} \circ f_0$, where π_1 denotes the blow-up we described, and reiterate the process.

We shall denote by $(x_0, y_0, t_0) := (x, y, t)$ the coordinates in 0 when considered in the image of $f_0 : (\mathbb{C}^3, 0) \to (\mathbb{C}^3, 0)$.

We shall also make all the computations of blow-ups in suitable local charts.

(1st **blow-up**). We blow-up the point $(0, 0, 0)$; we consider local coordinates (x_1, y_1, t_1) such that $(x_0, y_0, t_0) = \pi_1(x_1, y_1, t_1) = (x_1, x_1y_1, x_1t_1)$. Then the exceptional divisor is given by $E_1^{(1)} = \{x_1 = 0\}$, and the lifted map f_1 is given by

$$\begin{aligned} f_1(x, y, t) &:= \pi_1^{-1} \circ f_0(x, y, t) \\ &= \left(\frac{t^2(1 + yt)}{(1 + yt)^2 + ct^2 + xt^3}, \frac{t}{(1 + yt)}, \frac{t^2}{(1 + yt)}\right). \end{aligned}$$

We have that $f_1(t=0)=(0,0,0)$.

(2nd **blow-up**). We blow-up the point $(0,0,0)$; we consider local coordinates (x_2, y_2, t_2) such that $(x_1, y_1, t_1) = \pi_2(x_2, y_2, t_2) = (x_2y_2, y_2, y_2t_2)$. Then the exceptional divisor is given by $E_2^{(2)} = \{y_2 = 0\}$, and the lifted map f_2 is given by

$$f_2(x,y,t) := \pi_2^{-1} \circ f_1(x,y,t) = \left(\frac{t(1+yt)^2}{(1+yt)^2+ct^2+xt^3}, \frac{t}{(1+yt)}, t \right).$$

We have that $f_2(t=0)=(0,0,0)$.

(3rd **blow-up**). We blow-up the point $(0,0,0)$; we consider local coordinates (x_3, y_3, t_3) such that $(x_2, y_2, t_2) = \pi_3(x_3, y_3, t_3) = (x_3t_3, y_3t_3, t_3)$. Then the exceptional divisor is given by $E_3^{(3)} = \{t_3 = 0\}$, and the lifted map f_3 is given by

$$f_3(x,y,t) := \pi_3^{-1} \circ f_2(x,y,t) = \left(\frac{(1+yt)^2}{(1+yt)^2+ct^2+xt^3}, \frac{1}{(1+yt)}, t \right).$$

We have that $f_3(t=0)=(1,1,0)$.

(**Translation to** $(1,1,0)$). Before performing the next blow-up, we need to translate; we then consider local coordinates $(\widetilde{x}_3, \widetilde{y}_3, \widetilde{t}_3)$ such that $(x_3, y_3, t_3) = \widetilde{\tau}_3(\widetilde{x}_3, \widetilde{y}_3, \widetilde{t}_3) = (\widetilde{x}_3+1, \widetilde{y}_3+1, \widetilde{t}_3)$. Then the shifted map $\widetilde{f}_3$ is given by

$$\widetilde{f}_3(x,y,t) := \widetilde{\tau}_3^{-1} \circ f_3(x,y,t) = \left(\frac{-t^2(c+xt)}{(1+yt)^2+ct^2+xt^3}, \frac{-yt}{(1+yt)}, t \right).$$

Obviously we have $\widetilde{f}_3(t=0)=(0,0,0)$.

(4th **blow-up**). We blow-up the point $(0,0,0)$; we consider local coordinates (x_4, y_4, t_4) such that $(\widetilde{x}_3, \widetilde{y}_3, \widetilde{t}_3) = \widetilde{\pi}_4(x_4, y_4, t_4) = (x_4t_4, y_4t_4, t_4)$. Then the exceptional divisor is given by $E_4^{(4)} = \{t_4 = 0\}$, and the lifted map f_4 is given by

$$f_4(x,y,t) := \widetilde{\pi}_4^{-1} \circ \widetilde{f}_3(x,y,t) = \left(\frac{-t(c+xt)}{(1+yt)^2+ct^2+xt^3}, \frac{-y}{(1+yt)}, t \right).$$

We have that $f_4(x,y,0)=(0,-y,0)$, *i.e.*, $f_4(t=0)=\{x_4=t_4=0\}$.

(5th **blow-up**). We blow-up the line $\{x_4=t_4=0\}$; we consider local coordinates (x_5, y_5, t_5) such that $(x_4, y_4, t_4) = \pi_5(x_5, y_5, t_5) = (x_5t_5, y_5, t_5)$. Then the exceptional divisor is given by $E_5^{(5)} = \{t_5 = 0\}$, and the lifted map f_5 is given by

$$f_5(x,y,t) := \pi_5^{-1} \circ f_4(x,y,t) = \left(\frac{-(c+xt)}{(1+yt)^2+ct^2+xt^3}, \frac{-y}{(1+yt)}, t \right).$$

We have that $f_5(x, y, 0) = (-c, -y, 0)$, *i.e.*, $f_5(t = 0) = \{x_5 = -c, t_5 = 0\}$.

(**Translation to** $(-c, 0, 0)$). We again need to translate before performing the next blow-up; we then consider local coordinates $(\widetilde{x}_5, \widetilde{y}_5, \widetilde{t}_5)$ such that $(x_5, y_5, t_5) = \widetilde{\tau}_5(\widetilde{x}_5, \widetilde{y}_5, \widetilde{t}_5) = (\widetilde{x}_5 - c, \widetilde{y}_5, \widetilde{t}_5)$. Then the shifted map $\widetilde{f}_5$ is given by

$$\begin{aligned}\widetilde{f}_5(x, y, t) &:= \widetilde{\tau}_5^{-1} \circ f_5(x, y, t)\\ &= \left(\frac{t(2cy - x + cy^2t + c^2t + cxt^2)}{(1 + yt)^2 + ct^2 + xt^3}, \frac{-y}{(1 + yt)}, t\right).\end{aligned}$$

Obviously we have $\widetilde{f}_5(x, y, 0) = (0, -y, 0)$, *i.e.*, $\widetilde{f}_5(t = 0) = \{\widetilde{x}_5 = \widetilde{t}_5 = 0\}$.

(6th **blow-up**). We blow-up the line $\{\widetilde{x}_5 = \widetilde{t}_5 = 0\}$; we consider local coordinates (x_6, y_6, t_6) such that $(\widetilde{x}_5, \widetilde{y}_5, \widetilde{t}_5) = \widetilde{\pi}_6(x_6, y_6, t_6) = (x_6t_6, y_6, t_6)$. Then the exceptional divisor is given by $E_6^{(6)} = \{t_6 = 0\}$, and the lifted map f_6 is given by

$$f_6(x, y, t) := \widetilde{\pi}_6^{-1} \circ \widetilde{f}_5 = \left(\frac{2cy - x + cy^2t + c^2t + cxt^2}{(1 + yt)^2 + ct^2 + xt^3}, \frac{-y}{(1 + yt)}, t\right).$$

We have at last that f_6 is an invertible germ, that takes p into $q := (x_6 = 0, y_6 = 0, t_6 = 0)$.

Definition 4.2.4. We set $\pi_i = \widetilde{\tau}_{i-1} \circ \widetilde{\pi}_i$ for $i = 4, 6$. We shall denote by X_0 the projective space $\mathbb{P}^3$, and by X_i for $i = 1, \ldots 6$ the total space of the blow-up of X_0 through $\pi_1 \circ \ldots \circ \pi_i$. We shall also denote the strict transform of $E_i^{(i)}$ through $\pi_{i+1} \circ \cdots \circ \pi_k$ by $E_i^{(k)}$ (for $0 \leq i < k \leq 6$), that is a divisor in X_k.

We have then decomposed $f = \pi \circ \sigma$, where $\pi = \pi_1 \circ \ldots \circ \pi_6$ is a composition of blow-ups of smooth centers, and $\sigma = f_6$ is an invertible germ from p to the free point q.

Remark 4.2.5. While the "minimal" resolution $f = \pi \circ \sigma$ of a strict attracting germ $f : (\mathbb{C}^d, 0) \to (\mathbb{C}^d, 0)$ is unique in dimension $d = 2$, in higher dimension there is not a concept of minimal resolution, and indeed several resolutions are possible.

For example, in the resolution we gave of f_0 defined by (4.4), instead of performing the 2nd blow-up π_2, we could proceed as following.

We first blow-up the line $\{x_1 = y_1 = 0\}$, considering local coordinates (x_2', y_2', t_2') such that $(x_1, y_1, t_1) = \pi_2'(x_2', y_2', t_2') = (x_2'y_2', y_2', t_2')$.

Then we blow-up the line $\{y_2' = t_2' = 0\}$, considering local coordinates (x_3', y_3', t_3') such that $(x_2', y_2', t_2') = \pi_3'(x_3', y_3', t_3') = (x_3', y_3', y_3' t_3')$.

Then, locally with respect to the coordinates chosen, we have $\pi_2 = \pi_2' \circ \pi_3'$, and hence we could consider these two blow-ups instead of the original π_2, without changing the rest of the resolution, and we get a new resolution (made by seven blow-ups instead of six).

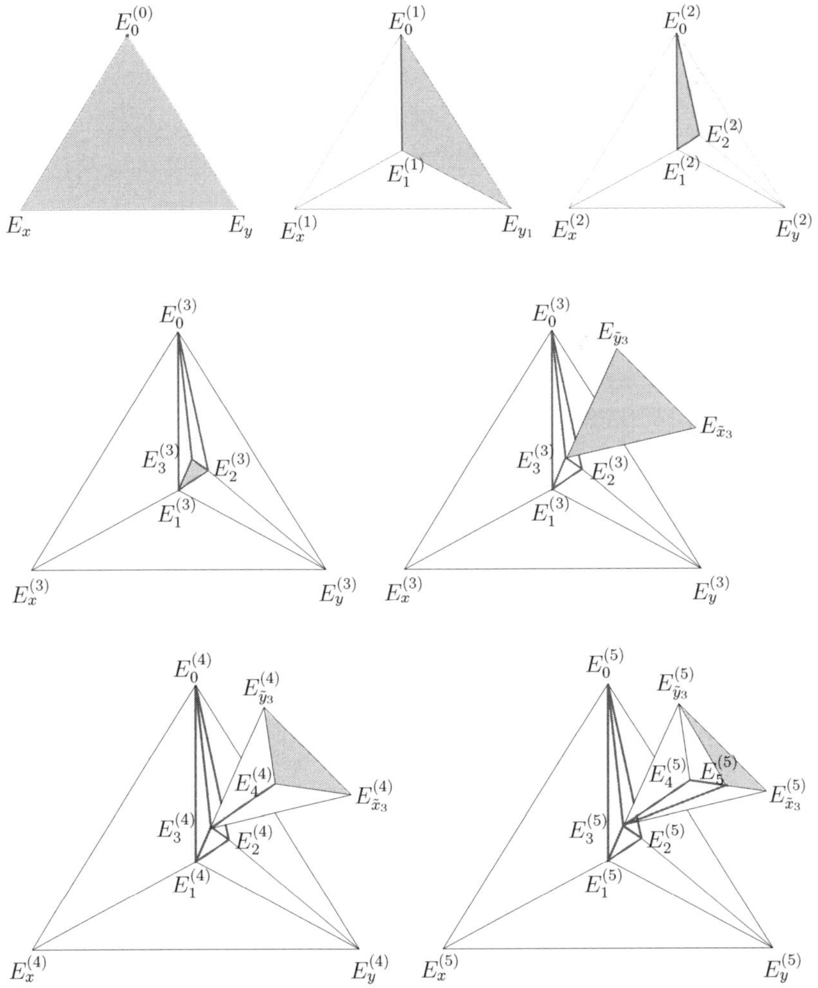

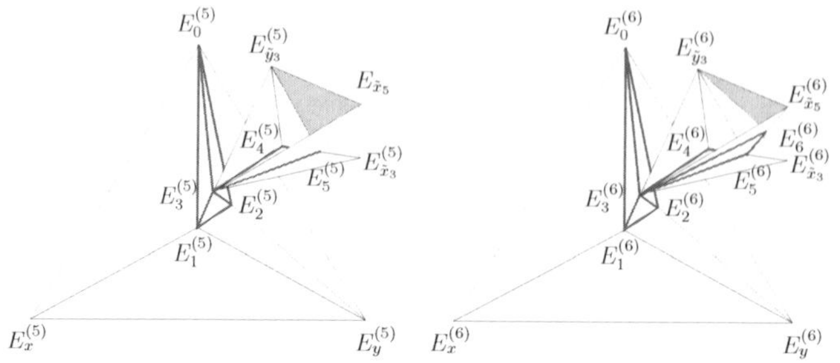

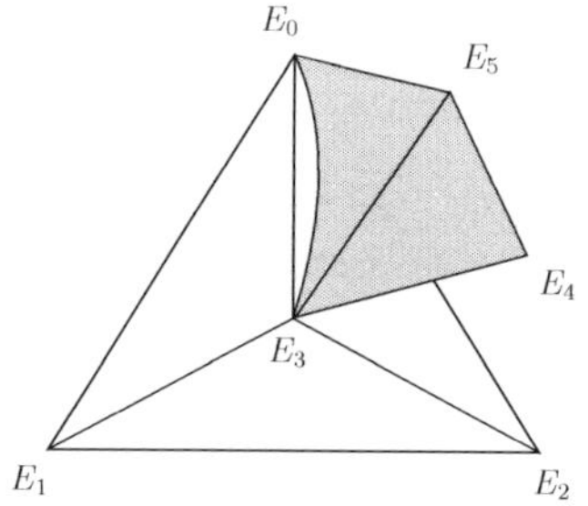

4.2.3. Construction and universal covering

The Kato variety

Let us consider the chart of $\mathbb{P}^3$ where the map f is given by f_0, centered in $p = [0:0:1:0]$.

We shall denote by $\mathcal{B}_\varepsilon$ the open ball of radius ε centered in 0, and by Σ_ε its boundary.

In Subsection 4.2.2 we proved that $f_0 = \pi \circ \sigma$, where $\pi = \pi_1 \circ \ldots \circ \pi_6$ and $\sigma = f_6$ is an invertible germ from p to $q \in E_6^{(6)}$; in particular $\sigma : \mathcal{B}_\varepsilon \to \sigma(\mathcal{B}_\varepsilon)$ is a biholomorphism for ε small enough.

We denote $\mathcal{B}_\varepsilon^\pi = \pi^{-1}(\mathcal{B}_\varepsilon)$, that is a neighborhood of the exceptional divisor $E = \bigcup_{i=1}^6 E_i^{(6)}$ of π, and $\Sigma_\varepsilon^\pi = \pi^{-1}(\Sigma_\varepsilon)$ its boundary, that intersects $E_0^{(6)}$.

On the other hand, $\sigma(\mathcal{B}_\varepsilon)$ is a neighborhood of q that intersects E only in $E_6^{(6)}$ for ε small enough.

We want to construct a 3-dimensional compact complex manifold, by taking $\mathcal{A}_\varepsilon^\pi := \overline{\mathcal{B}_\varepsilon^\pi} \setminus \sigma(\mathcal{B}_\varepsilon)$ and identifying Σ_ε^π and $\sigma(\Sigma_\varepsilon)$ through $\sigma \circ \pi$.

In fact, it is better to identify a neighborhood of Σ_ε^π and one of $\sigma(\Sigma_\varepsilon)$ through $\sigma \circ \pi$.

We set $\Sigma_{\varepsilon_-,\varepsilon_+} := \mathcal{B}_{\varepsilon_+} \setminus \overline{\mathcal{B}_{\varepsilon_-}}$, and denote $\Sigma_{\varepsilon_-,\varepsilon_+}^\pi = \pi^{-1}(\Sigma_{\varepsilon_-,\varepsilon_+}) =$

$\mathcal{B}^{\pi}_{\varepsilon_+} \setminus \overline{\mathcal{B}^{\pi}_{\varepsilon_-}}$; finally we consider

$$A := \mathcal{A}^{\pi}_{\varepsilon_-,\varepsilon_+} := \overline{\mathcal{B}^{\pi}_{\varepsilon_+}} \setminus \sigma(\mathcal{B}_{\varepsilon_-}). \tag{4.6}$$

For $0 < \varepsilon_- < \varepsilon_+$ small enough, we have that $\sigma \circ \pi$ is a biholomorphism between $\Sigma^{\pi}_{\varepsilon_-,\varepsilon_+}$ and $\sigma(\Sigma_{\varepsilon_-,\varepsilon_+})$. We finally define

$$X := A/\!\sim,$$

where $\sim$ denotes the identification given by $\sigma \circ \pi$ between $\Sigma^{\pi}_{\varepsilon_-,\varepsilon_+}$ and $\sigma(\Sigma_{\varepsilon_-,\varepsilon_+})$.

Since $\sigma \circ \pi : \Sigma^{\pi}_{\varepsilon_-,\varepsilon_+} \to \sigma(\Sigma_{\varepsilon_-,\varepsilon_+})$ is a biholomorphism, then X is a compact complex manifold. Moreover we have that $\sigma \circ \pi$ sends $\Sigma^{\pi}_{\varepsilon_-,\varepsilon_+} \cap E^{(6)}_0$ biholomorphically into $\sigma(\Sigma_{\varepsilon_-,\varepsilon_+}) \cap E^{(6)}_6$, and hence $E^{(6)}_0$ and $E^{(6)}_6$ glue together to get a divisor E_0.

Definition 4.2.6. Let X be the Kato variety described above. We shall denote by E_0 the divisor in X obtained gluing together $E^{(6)}_0$ and $E^{(6)}_6$. We shall moreover denote by E_i the divisor induced by $E^{(6)}_i$ in X for $i = 1, \dots, 5$.

Remark 4.2.7. As we have pointed in Remark 4.1.2, X admits a global spherical shell, the one given by the map $\sigma : \Sigma_{\varepsilon_-,\varepsilon_+} \to X$ (or equivalently $\pi^{-1} : \Sigma_{\varepsilon_-,\varepsilon_+} \to X$), well defined since we are identifying A through $\sigma \circ \pi$ to obtain X.

While the existence of a global spherical shell is preserved by blow-ups in dimension 2 (since we can blow-up only points, and we can move a GSS a little in order to avoid the blown-up point), this is not anymore true in higher dimensions, since we could blow-up a curve that intersects the GSS (even for small perturbations of it), loosing in this way the spherical shell.

So admitting a GSS is not a birational invariant, *i.e.*, even if X admits a GSS, there could be birationally equivalent models that do not admit any GSS.

The universal covering

We want now to construct a (universal) covering of X.

Consider $A = \mathcal{A}^{\pi}_{\varepsilon_-,\varepsilon_+}$ as in (4.6), and take $(A_i)_{i\in\mathbb{Z}}$ infinite copies of A. Instead of identifying through $\sigma \circ \pi$ the borders of the same copy, we shall identify $\Sigma^{\pi}_{\varepsilon_-,\varepsilon_+}$ in A_{i+1} with $\sigma(\Sigma_{\varepsilon_-,\varepsilon_+})$ in A_i for every $i \in \mathbb{Z}$. We get in this way a complex analytic (non-compact) 3-fold $\widetilde{X}$. Denoting by $(x_i)_{i\in\mathbb{Z}}$ the infinite copies of a point $x \in A$, we have then a natural automorphism

$$\widetilde{g} : \widetilde{X} \to \widetilde{X}$$

given by

$$\widetilde{g}(x_i) = x_{i+1}$$

on A_i. This automorphism is indeed well defined: if $y_i \sim x_{i+1}$, *i.e.*, $y_i = \sigma \circ \pi(x_{i+1})$, where $x_{i+1} \in \Sigma^{\pi}_{\varepsilon_-,\varepsilon_+} \subset A_{i+1}$ and $y_i \in \sigma(\Sigma_{\varepsilon_-,\varepsilon_+}) \subset A_i$, then

$$\widetilde{g}(y_i) = y_{i+1} = \sigma \circ \pi(x_{i+2}) = \sigma \circ \pi\big(\widetilde{g}(x_{i+1})\big),$$

and hence $\widetilde{g}(y_i) \sim \widetilde{g}(x_{i+1})$.

Moreover, the canonical projection

$$\omega : \widetilde{X} \to \widetilde{X}/\{\widetilde{g}^j \mid j \in \mathbb{Z}\} \cong X$$

makes $(\widetilde{X}, \omega)$ a covering of X.

We shall see that actually $\widetilde{X}$ is an universal covering (see Corollary 4.3.9).

Other constructions

Remark 4.2.8. Let us suppose to start from $X_0 = \mathbb{P}^3$ and to perform the 6 blow-ups $\pi_1, \ldots, \pi_6$ defined as in Definition 4.2.4, obtaining X_6. We showed that there is a biholomorphism $\sigma : (X_0, p) \to (X_6, q)$, and we glued together suitable neighborhoods of p and q to get our Kato variety X. Instead of gluing in this way, we can glue X_6 with A_1 as in (4.6), as we did for the universal covering. This means that, after the gluing, we perform another 6 blow-ups, $\pi_7, \ldots, \pi_{12}$, that are locally the same as $\pi_1, \ldots, \pi_6$, after the identification given by σ, obtaining some complex manifolds that we shall call respectively $X_7, \ldots, X_{12}$. We could obviously continue this kind of construction with an infinite number of copies $\{A_j\}_{j\in\mathbb{N}^*}$ of A, as we did for the universal covering, obtaining π_{i+6j} and X_{i+6j} for $i = 1, \ldots, 6$ and $j \in \mathbb{N}^*$; we can also take the limit of this process and get a non-compact manifold X_∞.

Definition 4.2.9. Let $X_0 = \mathbb{P}^3, \pi = \pi_1 \circ \ldots \circ \pi_6$ and $E_i^{(j)}$ for $0 \leq i \leq j \leq 6$ as in Definition 4.2.4, and $\sigma : (X_0, p) \to (X_6, q)$, where $f_0 = \pi \circ \sigma$ is the resolution of f_0 the germ as in (4.4), as we have seen in Subsection 4.2.2. We shall denote by π_k the blow-ups defined as in Remark 4.2.8 for $k \in \mathbb{N}^*$, and recursively, by X_k the blow-up of X_{k-1} through π_k. We shall define recursively on X_k, by $E_h^{(k)}$ the strict transform of $E_h^{(k-1)}$ through π_k for $h = 0, \ldots, k-1$, and by $E_k^{(k)}$ the exceptional divisor of π_k.

We shall denote by X_∞ the non-compact manifold obtained by taking the limit of this process.

Remark 4.2.10. We can define a map $g_j : A_j \to A_{j+1}$ as for the universal covering, that gives us isomorphisms between suitable neighborhoods of $E_i^{(k)}$ and $E_{i+6}^{(k)}$ if k is big enough, and $i \geq 1$. This is trivial for $6 \nmid i$, and it works in general since $E_0^{(6)} \cap E_6^{(6)}$ is empty. Moreover, locally there is a projection ω as for the universal covering, that gives an isomorphism between suitable neighborhoods of $E_{i+6j}^{(k)}$ and E_i in X, again for k big enough, a suitable j and $i = 1, \ldots, j$.

4.3. Algebraic properties

4.3.1. Topology of the divisors

We want now to explicit the topology of all these exceptional divisors.

Definition 4.3.1. We shall use the following notations.

(i) $\mathbb{P}^2$ denotes the complex projective plane and $D_0^{(0)}$ a line in it.

(ii) $\mathbb{P}_1^2$ is the blow-up of $\mathbb{P}^2$ at a (free) point in $D_0^{(0)}$, $D_0^{(1)}$ is the strict transform of $D_0^{(0)}$, while $D_1^{(1)}$ is the exceptional divisor.

(iii) $\mathbb{P}_2^2$ is the blow-up of $\mathbb{P}_1^2$ at the (free) point in $D_0^{(1)} \cap D_1^{(1)}$, $D_i^{(2)}$ denotes the strict transform of $D_i^{(1)}$ for $i = 0, 1$, and $D_2^{(2)}$ is the exceptional divisor.

(iv) $\mathbb{P}_{3a}^2$ is the blow-up of $\mathbb{P}_2^2$ at the (satellite) point in $D_1^{(2)} \cap D_2^{(2)}$, $D_i^{(3a)}$ denotes the strict transform of $D_i^{(2)}$ for $i = 0, 1, 2$, and $D_3^{(3a)}$ is the exceptional divisor.

(v) $\mathbb{P}_{3b}^2$ is the blow-up of $\mathbb{P}_2^2$ at the (free) point in $D_0^{(2)} \cap D_2^{(2)}$, $D_i^{(3b)}$ denotes the strict transform of $D_i^{(2)}$ for $i = 0, 1, 2$, and $D_3^{(3b)}$ is the exceptional divisor.

(vi) $\mathbb{F}_n$ is the ruled surfaces over $\mathbb{P}^1$ which has a curve L_∞ of self-intersection $-n$. We shall denote by H the fiber over a point.

(vii) We shall denote by $(\mathbb{F}_n)_1$ the blow-up of $\mathbb{F}_n$ at a point q in H but not in L_∞. We shall denote by $H^{(1)}$ the strict transform of H and by $D_1^{(1)}$ the exceptional divisor. We shall denote by $(\mathbb{F}_n)_2$ the blow-up of $(\mathbb{F}_n)_1$ at the (free) point in $H^{(1)} \cap D_1^{(1)}$. We shall denote by $H^{(2)}$ the strict transform of $H^{(1)}$, by $D_1^{(2)}$ the strict transform of $D_1^{(1)}$, and by $D_2^{(2)}$ the exceptional divisor. Finally, we shall denote by $(\mathbb{F}_n)_{3a}$ the blow-up of $(\mathbb{F}_n)_2$ at the (satellite) point in $D_1^{(2)} \cap D_2^{(2)}$. We shall denote by $H^{(3a)}$ the strict transform of $H^{(2)}$, by $D_i^{(3)}$ the strict transform of $D_i^{(2)}$ for $i = 1, 2$, and by $D_3^{(3a)}$ the exceptional divisor.

In every case, with an abuse of notation, we shall denote by a divisor D also its pull-back by a suitable modification. For example, we shall denote by H in $(\mathbb{F}_n)_1$, $(\mathbb{F}_n)_2$ and $(\mathbb{F}_n)_{3a}$ respectively, the divisor

$$H = H^{(1)} + D_1^{(1)} = H^{(2)} + D_1^{(2)} + 2D_2^{(2)} = H^{(3)} + D_1^{(3)} + 2D_2^{(3)} + 3D_3^{(3)}.$$

Proposition 4.3.2. *Let X be the Kato variety and $E_0, \ldots, E_5$ the divisors described in Definition* 4.2.6. *Then*

$$
\begin{aligned}
E_0 &\cong (\mathbb{F}_3)_{3a},\\
E_1 &\cong \mathbb{P}^2_2,\\
E_2 &\cong \mathbb{P}^2_1,\\
E_3 &\cong \mathbb{P}^2_{3b},\\
E_4 &\cong \mathbb{P}^2,\\
E_5 &\cong \mathbb{F}_2.
\end{aligned}
$$

Proof. We shall proceed step-by-step; we start from $E_0^{(0)} \cong \mathbb{P}^2$ (actually we are interested on the local structure in p, so we can think of $E_0^{(0)}$ as an open ball in $\mathbb{C}^2$).

(1st **blow-up**). We blow-up a point in $E_0^{(0)}$, so we have

$$E_0^{(1)} \cong \mathbb{P}^2_1, \qquad E_1^{(1)} \cong \mathbb{P}^2.$$

(2nd **blow-up**). We blow-up a point in the intersection of $E_0^{(1)}$ and $E_1^{(1)}$, so we have

$$E_0^{(2)} \cong \mathbb{P}^2_2, \qquad E_1^{(2)} \cong \mathbb{P}^2_1, \qquad E_2^{(2)} \cong \mathbb{P}^2.$$

(3rd **blow-up**). We blow-up the intersection point between $E_0^{(2)}$, $E_1^{(2)}$ and $E_2^{(2)}$, so we have

$$E_0^{(3)} \cong \mathbb{P}^2_{3a}, \qquad E_1^{(3)} \cong \mathbb{P}^2_2, \qquad E_2^{(3)} \cong \mathbb{P}^2_1, \qquad E_3^{(3)} \cong \mathbb{P}^2.$$

(4th **blow-up**). We blow-up a free point in $E_3^{(3)}$ (*i.e.*, this point doesn't belong to other exceptional divisors), so we have

$$E_0^{(4)} \cong \mathbb{P}^2_{3a}, \quad E_1^{(4)} \cong \mathbb{P}^2_2, \quad E_2^{(4)} \cong \mathbb{P}^2_1, \quad E_3^{(4)} \cong \mathbb{P}^2_1, \quad E_4^{(4)} \cong \mathbb{P}^2.$$

(5th **blow-up**). We blow-up a line L in $E_4^{(4)} \cong \mathbb{P}^2$ that intersects $E_3^{(4)}$ in one point, so we have

$$E_i^{(5)} \cong E_i^{(4)} \text{ for } i = 0, 1, 2, 4, \qquad E_3^{(5)} \cong \mathbb{P}^2_2, \qquad E_5^{(5)} \cong \mathbb{F}_2.$$

The last relation is obtained since we are blowing up the rational curve L, and hence $E_5^{(5)}$ is a rational ruled surface. Moreover $\mathcal{N}_{L\subset E_4^{(4)}} = \mathcal{O}(1)$ and $\mathcal{N}_{E_4^{(4)}\subset X_4}\Big|_L = \mathcal{O}(-1)$ (see Subsection 4.3.5(5th blow-up)). Hence, by Corollary 1.4.25, $E_5^{(5)} \cong \mathbb{F}_{1,-1} \cong \mathbb{F}_2$.

(6th **blow-up**). We blow-up a curve C in $E_5^{(5)} \cong \mathbb{F}_2$ that intersects $E_3^{(5)}$ in one point ($C = L_{-c}$ with the notations of Remark 1.4.24), so we have

$$E_i^{(6)} \cong E_i^{(5)} \text{ for } i = 0, 1, 2, 4, 5, \qquad E_3^{(6)} \cong \mathbb{P}^2_{3b}, \qquad \text{and } E_6^{(6)} \cong \mathbb{F}_3.$$

The last relation is obtained since we are blowing up the rational curve C, and hence $E_6^{(6)}$ is a rational ruled surface. Moreover $\mathcal{N}_{C\subset E_5^{(5)}} = \mathcal{O}(2)$ (see Proposition 1.4.26) and $\mathcal{N}_{E_5^{(5)}\subset X_5}\Big|_C = \mathcal{O}(-1)$ (see Subsection 4.3.5(6th blow-up)). Hence, by Corollary 1.4.25, $E_6^{(6)} \cong \mathbb{F}_{2,-1} \cong \mathbb{F}_3$.

(**Gluing**). After the identification, the only divisors that changes their topology are the ones glued together. So

$$E_i \cong E_i^{(6)} \text{ for } i = 1, \ldots, 5.$$

The new divisor is made by $E_6^{(6)}$, where we replace a neighborhood of a free point (not in L_∞) with a neighborhood of the blown-up point. It follows that

$$E_0 \cong (\mathbb{F}_3)_{3a}.$$

We notice that since $y_6 = -y$ in $E_6^{(6)}$, the role of $D_0^{(0)}$ in the starting X_0 is played here by H, and hence the sequence of blow-ups is the one described in Definition 4.3.1. □

E	$E_\cdot^{(0)}$	$E_\cdot^{(1)}$	$E_\cdot^{(2)}$	$E_\cdot^{(3)}$	$E_\cdot^{(4)}$	$E_\cdot^{(5)}$	$E_\cdot^{(6)}$
$E_0^{(\cdot)}$	$\mathbb{P}^2$	$\mathbb{P}^2_1$	$\mathbb{P}^2_2$	$\mathbb{P}^2_{3a}$	$\mathbb{P}^2_{3a}$	$\mathbb{P}^2_{3a}$	$\mathbb{P}^2_{3a}$
$E_1^{(\cdot)}$		$\mathbb{P}^2$	$\mathbb{P}^2_1$	$\mathbb{P}^2_2$	$\mathbb{P}^2_2$	$\mathbb{P}^2_2$	$\mathbb{P}^2_2$
$E_2^{(\cdot)}$			$\mathbb{P}^2$	$\mathbb{P}^2_1$	$\mathbb{P}^2_1$	$\mathbb{P}^2_1$	$\mathbb{P}^2_1$
$E_3^{(\cdot)}$				$\mathbb{P}^2$	$\mathbb{P}^2_1$	$\mathbb{P}^2_2$	$\mathbb{P}^2_{3b}$
$E_4^{(\cdot)}$					$\mathbb{P}^2$	$\mathbb{P}^2$	$\mathbb{P}^2$
$E_5^{(\cdot)}$						$\mathbb{F}_2$	$\mathbb{F}_2$
$E_6^{(\cdot)}$							$\mathbb{F}_3$

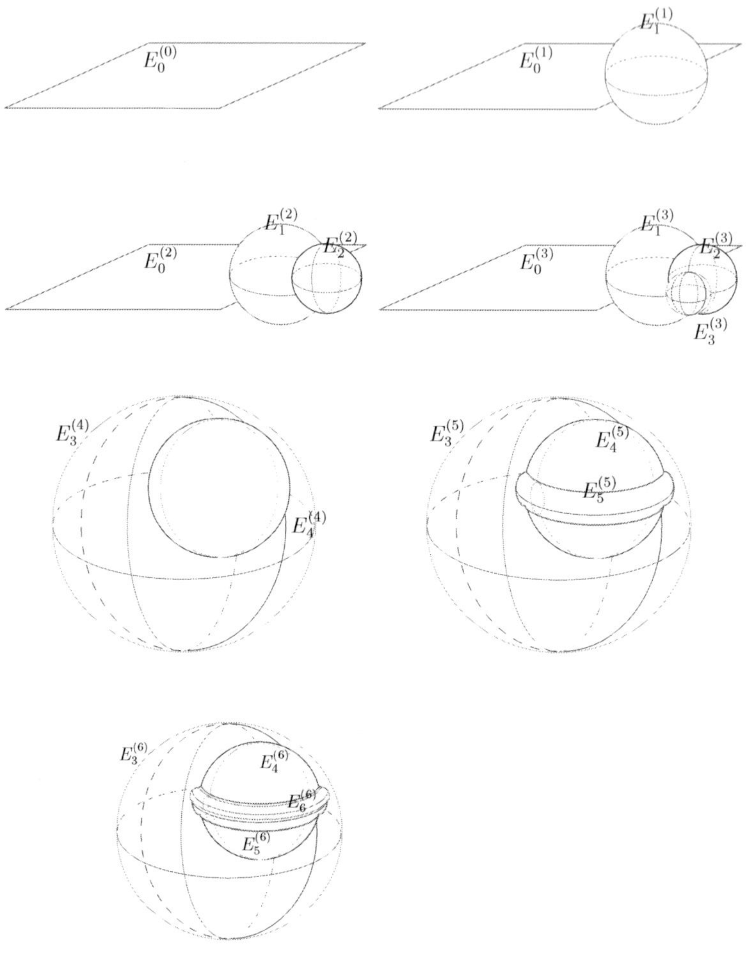
$E_0^{(0)}$
$E_1^{(1)}$
$E_0^{(1)}$
$E_1^{(2)}$
$E_2^{(2)}$
$E_0^{(2)}$
$E_1^{(3)}$
$E_2^{(3)}$
$E_0^{(3)}$
$E_3^{(3)}$
$E_3^{(4)}$
$E_4^{(4)}$
$E_3^{(5)}$
$E_4^{(5)}$
$E_5^{(5)}$
$E_3^{(6)}$
$E_4^{(6)}$
$E_6^{(6)}$
$E_5^{(6)}$

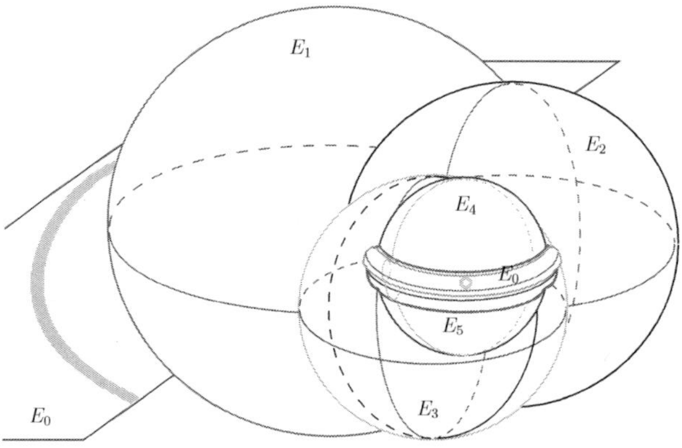
E_1
E_2
E_4
E_0
E_5
E_3
E_0

4.3.2. Homology Groups

In this subsection we will compute the homology groups of our manifold.

Proposition 4.3.3 ([31, p. 605]). *Let X be a compact complex manifold, and $\pi : \widetilde{X} \to X$ the blow-up of a submanifold L in X. Let us denote by E the exceptional divisor. Then we have*

$$H^*(\widetilde{X}) = \pi^* H^*(X) \oplus \big(H^*(E)/\pi^* H^*(L)\big).$$

Corollary 4.3.4. *Let X be a compact complex manifold of dimension n, and $\pi : \widetilde{X} \to X$ the blow-up of a point p in X. Let us denote by E the exceptional divisor. Then we have*

$$H^i(\widetilde{X}) = \pi^* H^i(X) \oplus \begin{cases} \mathbb{Z} & \text{if } i = 2k,\ k = 1, \dots, n-1, \\ 0 & \text{otherwise.} \end{cases}$$

Proof. We apply Proposition 4.3.3 to $L = \{p\}$; here $E \cong \mathbb{P}^{n-1}$, for which the cohomology is

$$H^i(\mathbb{P}^{n-1}) = \begin{cases} \mathbb{Z} & \text{if } i = 2k,\ k = 0, \dots, n-1, \\ 0 & \text{otherwise.} \end{cases}$$

Moreover $H^i(L) = \mathbb{Z}$ if $i = 0$ and it is trivial otherwise. Finally $\pi^* H^0(L) \cong \mathbb{Z}$ and we are done. □

Proposition 4.3.5. *The homology groups for $E = \mathbb{F}_h$ are*

$$H_0(E) = H_4(E) = \mathbb{Z}, \qquad H_2(E) = \mathbb{Z}^2, \qquad H_i(E) = 0 \text{ otherwise.}$$

Proof. E has a structure of bundle $\pi : E \to L$ over $L \cong \mathbb{P}^1$, with fiber $\mathbb{P}^1$. We denote $L \cong \mathbb{P}^1 = \mathbb{C} \cup \{\infty\}$. We want to use twice Theorem 1.3.5 (Mayer-Vietoris sequence in homology). Let us consider $U = \pi^{-1}(\mathbb{C})$ and $V = E \setminus L$ where L here denotes the 0-section of the bundle. In this case, $U \sim \mathbb{P}^1$ (being $\mathbb{C}$ contractible), and $V \sim \mathbb{P}^1$ (we can retract V to the ∞-section for example), $U \cap V \cong \mathbb{C}^2 \sim \{p\}$, and $U \cup V = E \setminus \{q\} =: \hat{E}$, where q is the point ∞ in the 0-section L. Then Theorem 1.3.5 applied to U and V is

$$\begin{array}{cccccccccc}
\to & 0 & \to & 0 & \oplus & 0 & \to & H_4(\widehat{E}) & \to & \\
\to & 0 & \to & 0 & \oplus & 0 & \to & H_3(\widehat{E}) & \to & \\
\to & 0 & \to & \mathbb{Z} & \oplus & \mathbb{Z} & \to & H_2(\widehat{E}) & \to & \\
\to & 0 & \to & 0 & \oplus & 0 & \to & H_1(\widehat{E}) & \to & \\
\to & 0 & \to & 0 & \oplus & 0 & \to & \widetilde{H}_0(\widehat{E}) & \to & 0,
\end{array}$$

and hence $H_2(\widehat{E}) \cong \mathbb{Z} \oplus \mathbb{Z} \cong \mathbb{Z}^2$, and $\widetilde{H}_i(\widehat{E}) = 0$ otherwise.

We now use Theorem 1.3.5 for $\widehat{E}$ and B, where B is a small ball with center q. Here $\widehat{E} \cap B \sim \mathbb{S}^3$, while $\widehat{E} \cup B = E$. Then we get

$$\begin{array}{ccccccccccc}
\to & 0 & \to & 0 & \oplus & 0 & \to & H_4(E) & \to & \\
\to & \mathbb{Z} & \to & 0 & \oplus & 0 & \to & H_3(E) & \to & \\
\to & 0 & \to & \mathbb{Z}^2 & \oplus & 0 & \to & H_2(E) & \to & \\
\to & 0 & \to & 0 & \oplus & 0 & \to & H_1(E) & \to & \\
\to & 0 & \to & 0 & \oplus & 0 & \to & \widetilde{H}_0(E) & \to & 0,
\end{array}$$

and hence $H_2(E) \cong \mathbb{Z}^2$, $H_0(E) \cong H_4(E) \cong \mathbb{Z}$ and $H_i(E) = 0$ otherwise. □

Corollary 4.3.6. *Let X be a compact complex manifold of dimension 3, and $\pi : \widetilde{X} \to X$ the blow-up of a line $L \cong \mathbb{P}^1$ in X. Then we have*

$$H^i(\widetilde{X}) = \pi^* H^i(X) \oplus \begin{cases} \mathbb{Z} & \textit{if } i = 2, 4, \\ 0 & \textit{otherwise.} \end{cases}$$

Proof. We apply Proposition 4.3.3, using the computation of Proposition 4.3.5. We have $H^i(L) = H^i(\mathbb{P}^1) = \mathbb{Z}$ if $i = 0, 2$, and $H^i(L) = 0$ otherwise. Moreover we have

$$\begin{aligned} \pi^* : H^0(L) &\to H^0(E) \\ 1 &\mapsto 1, \\ \pi^* : H^2(L) &\to H^2(E) \\ 1 &\mapsto (1, 0), \end{aligned}$$

and we get the statement. □

Corollary 4.3.7. *For $\varepsilon > 0$ small enough, we have for $\mathcal{B}^\pi = \mathcal{B}^\pi_\varepsilon$ constructed as in Subsection 4.2.3 that*

$$\begin{array}{lll} H_0(\mathcal{B}^\pi) \cong \mathbb{Z}, & H_2(\mathcal{B}^\pi) \cong \mathbb{Z}^6, & H_4(\mathcal{B}^\pi) \cong \mathbb{Z}^6, \\ H_6(\mathcal{B}^\pi) \cong \mathbb{Z}, & H_i(\mathcal{B}^\pi) = 0 \quad \textit{otherwise.} & \end{array}$$

Proof. It follows directly from Corollaries 4.3.4 and 4.3.6 and Theorem 1.3.7 (Poincaré Duality). □

Proposition 4.3.8. *For $0 < \varepsilon_- < \varepsilon_+$ small enough, we have for $X = \mathcal{A}^\pi_{\varepsilon_-,\varepsilon_+}/\!\sim$ constructed as in Subsection 4.2.3 that*

$$\begin{array}{llll} H_0(X) \cong \mathbb{Z}, & H_1(X) \cong \mathbb{Z}, & H_2(X) \cong \mathbb{Z}^6, & H_3(X) = 0, \\ H_4(X) \cong \mathbb{Z}^6, & H_5(X) \cong \mathbb{Z}, & H_6(X) \cong \mathbb{Z}. & \end{array}$$

Proof. Let us set $\varepsilon = (\varepsilon_+ + \varepsilon_-)/2$ and $\delta = (\varepsilon_+ - \varepsilon_-)/2$, and denote by pr the projection pr : $\mathcal{A}^\pi_{\varepsilon_-,\varepsilon_+} \to X$. We want to use twice Theorem 1.3.5 (Mayer-Vietoris sequence in homology). First we consider Theorem 1.3.5 for $U = \mathcal{A}^\pi_{\varepsilon,\varepsilon} = \overset{\circ}{\mathcal{A}}{}^\pi_\varepsilon$ and $V = \sigma(\mathcal{B}_{\epsilon_+})$.

We have $V \cong \mathbb{B}^6 \sim \{p\}$, while $U \cap V = \sigma(\Sigma_{\varepsilon,\varepsilon_+}) \sim \mathbb{S}^5$ and $U \cup V = \mathcal{B}^\pi_{\varepsilon_+}$, for which we already computed the homology group in Corollary 4.3.7. Then the Mayer-Vietoris sequence is

$$\begin{array}{ccccccccc}
\to & 0 & \to & H_4(U) & \oplus & 0 & \to & \mathbb{Z}^6 & \to \\
\to & 0 & \to & H_3(U) & \oplus & 0 & \to & 0 & \to \\
\to & 0 & \to & H_2(U) & \oplus & 0 & \to & \mathbb{Z}^6 & \to \\
\to & 0 & \to & H_1(U) & \oplus & 0 & \to & 0 & \to \\
\to & 0 & \to & \widetilde{H}_0(U) & \oplus & 0 & \to & 0 & \to & 0.
\end{array}$$

Then we get $H_0(U) = \mathbb{Z}$, $H_1(U) = 0$, $H_2(U) = \mathbb{Z}^6$, $H_3(U) = 0$, $H_4(U) = \mathbb{Z}^6$.

We shall need also Theorem 1.3.6 (Mayer-Vietoris sequence in cohomology), getting

$$\begin{array}{ccccccccc}
0 & \to & 0 & \to & \widetilde{H}^0(U) & \oplus & 0 & \to & 0 & \to \\
 & \to & 0 & \to & H^1(U) & \oplus & 0 & \to & 0 & \to
\end{array}$$

and hence $\widetilde{H}^0(U) = 0$ and $H^1(U) = 0$.

Now we consider $\mathrm{pr}(U) \cong U$ and $W = \mathrm{pr}(\Sigma^\pi_{\varepsilon_-,\varepsilon_+})$: we have $\mathrm{pr}(U) \cup W = X$, $W \sim \mathbb{S}^5$ and $W \cap \mathrm{pr}(U) \sim \mathbb{S}^5 \sqcup \mathbb{S}^5$, where $\sqcup$ denotes the disjoint union. We have

$$\begin{array}{ccccccccc}
\to & 0 & \to & Z^6 & \oplus & 0 & \to & H_4(X) & \to \\
\to & 0 & \to & 0 & \oplus & 0 & \to & H_3(X) & \to \\
\to & 0 & \to & Z^6 & \oplus & 0 & \to & H_2(X) & \to \\
\to & 0 & \to & 0 & \oplus & 0 & \to & H_1(X) & \to \\
\to & \mathbb{Z} & \to & 0 & \oplus & 0 & \to & H_0(X) & \to & 0.
\end{array}$$

Then we get $H_0(X) = \mathbb{Z}$, $H_1(X) = \mathbb{Z}$, $H_2(X) = \mathbb{Z}^6$, $H_3(X) = 0$, $H_4(X) = \mathbb{Z}^6$, while, since X is a compact orientable manifold of dimension 6, we have $H_6(X) = \mathbb{Z}$ and $H_5(X) \cong H^1(X)$ thanks to Theorem 1.3.7 (Poincaré Duality).

Using Theorem 1.3.6, we get

$$\begin{array}{ccccccccc}
0 & \to & \widetilde{H}^0(X) & \to & 0 & \oplus & 0 & \to & \mathbb{Z} & \to \\
 & \to & H^1(X) & \to & 0 & \oplus & 0 & \to & 0 & \to
\end{array}$$

and hence $H_5(X) \cong H^1(X) = \mathbb{Z}$, and we are done. □

Corollary 4.3.9. *The covering $\widetilde{X}$ of X defined in Subsection 4.2.3 is an universal covering.*

Proof. Let A and $(A_i)_{i\in\mathbb{Z}}$ be as in Subsection 4.2.3. From the proof of Proposition 4.3.8 it follows that the first fundamental group $\pi_1(A) = 0$, while the sets $\Sigma^{\pi}_{\varepsilon_-,\varepsilon_+}$ and $\sigma(\Sigma_{\varepsilon_-,\varepsilon_+})$ are homotopically equivalent to $\mathbb{S}^5$, that is again simply connected. Then the assertion follows from Theorem 1.3.2 (Van Kampen Theorem). □

Corollary 4.3.10. *The Kato variety X defined as in Subsection* 4.2.3 *is non-Kähler.*

Proof. From Proposition 4.3.8 we have that the first Betti number of X is $b_1(X) = 1$, while for Kähler manifolds the odd Betti numbers have to be even (see 1.4.13). □

Proposition 4.3.11. *Let X and Y be two compact complex birationally equivalent* 3*-folds. Then the first fundamental groups $\pi_1(X)$ and $\pi_1(Y)$ are isomorphic.*

Proof. Without loss of generality, we can suppose that $Y = \widetilde{X}$ is the blow-up $\pi : Y \to X$ along a center C (a point or a curve).

Let us consider the induced map $\pi_* : \pi_1(\widetilde{X}) \to \pi_1(X)$.

Let us take an arc γ in X, and denote by $[\gamma]$ its class in $\pi_1(X)$. Since C has (real) dimension 2 in a 6-dimensional space, we can perturb γ continuously in order to avoid C. But π is invertible outside C, hence we can consider $\widetilde{\gamma} = \pi^{-1} \circ \gamma$, that satisfies $\pi_*([\widetilde{\gamma}]) = [\gamma]$, and π_* is surjective.

Analogously, we can consider $\widetilde{\gamma}$ an arc in $\widetilde{X}$, and perturb it in order to avoid the exceptional divisor E (since it has (real) dimension 4 in a 6-dimensional space); this implies the injectivity of π_*. □

Remark 4.3.12. Proposition 4.3.11 implies that the first Betti number b_1 is a birational invariant (in the sense that every birationally equivalent models have the same first Betti number). It follows that all birationally equivalent models of the Kato variety X we are studying have $b_1 = 1$ and hence they are all non-Kähler. For the other Betti numbers, we have $b_0 = b_6 = 1$ the number of connected components, while we have seen that $b_1 = b_5$ are birational invariants. On the contrary, $b_2 = b_4$ and b_3 can change under performing blow-ups. Indeed, we have seen that if we blow-up a point or a line, the second Betti number b_2 grows by one, while b_3 can change by blowing-up curves of genus greater than or equal to 1 (it can be shown with similar computations to the ones showed above in this section).

4.3.3. Intersection numbers

We want now to compute all the intersection numbers for all the divisors we have.

Definition 4.3.13. We shall use the following encoding. If we consider $E_i^{(k)}$ for $0 \leq i \leq k$ in X_k, we shall write

$$E_i^{(k)} : M_i,$$

where $M_i = (m_{i,j,h})_{j,h} \in \mathcal{M}(i \times i, \mathbb{Z})$ is such that $m_{i,j,h} = E_i^{(k)} \cdot E_j^{(k)} \cdot E_h^{(k)}$, and we shall write only entries with $j \leq h$ ($\leq i$). Since intersection numbers are invariant under permutation of the divisors, we get all the intersection numbers, and one entry for each triple up to permutations.

We shall use analogous notations for $E_0, \ldots E_5$ in X.

Lemma 4.3.14. *Let $\pi : Y \to X$ be a single blow-up of a point p in a 3-manifold X, and denote by E the exceptional divisor. Let $D_1, \ldots, D_h$ and $E_1, \ldots, E_k$ be divisors in X such that for all i, j we have $\pi^*(D_i) = \widetilde{D}_i$ and $\pi^*(E_j) = \widetilde{E}_j + E$, that is, $p \notin D_i$ while p is a smooth point in E_j, where $\widetilde{D}$ denotes the strict transform of a divisor D through π. Then we have the following relations.*

(i) $E^3 = 1$.
(ii) $E^2 \cdot \widetilde{E}_i = -1$, $E \cdot \widetilde{E}_i \cdot \widetilde{E}_j = 1$ *for all* i, j.
(iii) $E^2 \cdot \widetilde{D}_i = 0$, $E \cdot \widetilde{D}_i \cdot \widetilde{D}_j = 0$ *for all* i, j.
(iv) $E \cdot \widetilde{D}_i \cdot \widetilde{E}_j = 0$ *for all* i, j.
(v) $\widetilde{D}_i \cdot \widetilde{D}_j \cdot \widetilde{D}_l = D_i \cdot D_j \cdot D_l$, $\widetilde{D}_i \cdot \widetilde{D}_j \cdot \widetilde{E}_l = D_i \cdot D_j \cdot E_l$ *and* $\widetilde{D}_i \cdot \widetilde{E}_j \cdot \widetilde{E}_l = D_i \cdot E_j \cdot E_l$ *for all* i, j, l.
(vi) $\widetilde{E}_i \cdot \widetilde{E}_j \cdot \widetilde{E}_l = E_i \cdot E_j \cdot E_l - 1$ *for all* i, j, l.

Proof. First of all, recalling Remark 1.2.26, we can apply Propositions 1.2.24 and 1.2.25.

From Proposition 1.2.25 applied to E_i, E, E, we get

$$0 = E^3 + E^2 \cdot \widetilde{E}_i$$

and hence

$$E^2 \cdot \widetilde{E}_i = -E^3.$$

From Proposition 1.2.25 applied to E_i, E_j, E, we get

$$0 = E^3 + E^2 \cdot (\widetilde{E}_i + \widetilde{E}_j) + E \cdot \widetilde{E}_i \cdot \widetilde{E}_j$$

and hence

$$E \cdot \widetilde{E}_i \cdot \widetilde{E}_j = E^3.$$

From Proposition 1.2.24 applied to E_i, E_j, E_l, we get

$$\begin{aligned} E_i \cdot E_j \cdot E_l &= E^3 + E^2(\widetilde{E}_i + \widetilde{E}_j + \widetilde{E}_l) \\ &\quad + E(\widetilde{E}_i \cdot \widetilde{E}_j + \widetilde{E}_i \cdot \widetilde{E}_l + \widetilde{E}_j \cdot \widetilde{E}_l) + \widetilde{E}_i \cdot \widetilde{E}_j \cdot \widetilde{E}_l, \end{aligned}$$

and hence

$$\widetilde{E}_i \cdot \widetilde{E}_j \cdot \widetilde{E}_l = E_i \cdot E_j \cdot E_l - E^3.$$

If we pick in the last relation E_i, E_j, E_l three distinct surfaces that intersect transversely at p, then $E_i \cdot E_j \cdot E_l = 1$, while $\widetilde{E}_i$, $\widetilde{E}_j$, $\widetilde{E}_l$ do not intersect, and hence $\widetilde{E}_i \cdot \widetilde{E}_j \cdot \widetilde{E}_l = 0$ and $E^3 = 1$. So we got (i), (ii) and (vi). The point (iii) is given directly by Proposition 1.2.25. To prove (iv), we apply Proposition 1.2.25 to D_i, E_j, E: we get

$$0 = E^2 \cdot \widetilde{D}_i + E \cdot \widetilde{D}_i \cdot \widetilde{E}_j$$

and we conclude using (iii). Finally, to prove (v), the first case is given directly by Proposition 1.2.24. For the second case, we apply Proposition 1.2.24 to D_i, D_j, E_l: we get

$$D_i \cdot D_j \cdot E_l = \widetilde{D}_i \cdot \widetilde{D}_j \cdot E + \widetilde{D}_i \cdot \widetilde{D}_j \cdot \widetilde{E}_l,$$

and we conclude again by using (iii). The third case follows analogously using (iii) and (iv). □

Proposition 4.3.15. *Let X be the Kato variety and $E_0, \dots, E_5$ the divisors described in Definition 4.2.6. Then we have the following situation for intersection numbers.*

$$E_0 : \begin{pmatrix} -4 \end{pmatrix} \qquad E_1 : \begin{pmatrix} -1 & -3 \\ & -1 \end{pmatrix}$$

$$E_2 : \begin{pmatrix} 0 & 0 & -2 \\ & 0 & -2 \\ & & 0 \end{pmatrix} \qquad E_3 : \begin{pmatrix} 0 & 1 & 1 & -1 \\ & 1 & 1 & -1 \\ & & 1 & -1 \\ & & & 0 \end{pmatrix}$$

$$E_4 : \begin{pmatrix} 0 & 0 & 0 & 0 & 0 \\ & 0 & 0 & 0 & 0 \\ & & 0 & 0 & 0 \\ & & & 1 & -2 \\ & & & & 4 \end{pmatrix} \qquad E_5 : \begin{pmatrix} 2 & 0 & 0 & 1 & 0 & -3 \\ & 0 & 0 & 0 & 0 & 0 \\ & & 0 & 0 & 0 & 0 \\ & & & 0 & 1 & -2 \\ & & & & -2 & 1 \\ & & & & & 4 \end{pmatrix}.$$

Proof. We proceed step by step, blow-up by blow-up. We start from one divisor, $E_0^{(0)}$, which has self-intersection ${E_0^{(0)}}^3 = 1$ (being a plane in $\mathbb{P}^3$).

$$E_0^{(0)} : \begin{pmatrix} 1 \end{pmatrix}$$

(1st **blow-up**). We have $\pi_1^*(E_0^{(0)}) = E_0^{(1)} + E_1^{(1)}$. Thanks to Lemma 4.3.14 we have

$$E_0^{(1)} : \begin{pmatrix} 0 \end{pmatrix} \quad E_1^{(1)} : \begin{pmatrix} 1 & -1 \\ & 1 \end{pmatrix}.$$

(2nd **blow-up**). We have $\pi_2^*(E_0^{(1)}) = E_0^{(2)} + E_2^{(2)}$ and $\pi_2^*(E_1^{(1)}) = E_1^{(2)} + E_2^{(2)}$. Thanks to Lemma 4.3.14 we have

$$E_0^{(2)} : \begin{pmatrix} -1 \end{pmatrix} \quad E_1^{(2)} : \begin{pmatrix} 0 & -2 \\ & 0 \end{pmatrix} \quad E_2^{(2)} : \begin{pmatrix} 1 & 1 & -1 \\ & 1 & -1 \\ & & 1 \end{pmatrix}.$$

(3rd **blow-up**). We have $\pi_3^*(E_0^{(2)}) = E_0^{(3)} + E_3^{(3)}$, $\pi_3^*(E_1^{(2)}) = E_1^{(3)} + E_3^{(3)}$ and $\pi_3^*(E_2^{(2)}) = E_2^{(3)} + E_3^{(3)}$. Thanks to Lemma 4.3.14 we have

$$E_0^{(3)} : \begin{pmatrix} -2 \end{pmatrix} \quad E_1^{(3)} : \begin{pmatrix} -1 & -3 \\ & -1 \end{pmatrix} \quad E_2^{(3)} : \begin{pmatrix} 0 & 0 & -2 \\ & 0 & -2 \\ & & 0 \end{pmatrix}$$

$$E_3^{(3)} : \begin{pmatrix} 1 & 1 & 1 & -1 \\ & 1 & 1 & -1 \\ & & 1 & -1 \\ & & & 1 \end{pmatrix}.$$

(4th **blow-up**). We have $\pi_4^*(E_0^{(3)}) = E_0^{(4)}$, $\pi_4^*(E_1^{(3)}) = E_1^{(4)}$, $\pi_4^*(E_2^{(3)}) = E_2^{(4)}$ and $\pi_4^*(E_3^{(3)}) = E_3^{(4)} + E_4^{(4)}$. Thanks to Lemma 4.3.14 we have

$$E_0^{(4)} : \begin{pmatrix} -2 \end{pmatrix} \quad E_1^{(4)} : \begin{pmatrix} -1 & -3 \\ & -1 \end{pmatrix} \quad E_2^{(4)} : \begin{pmatrix} 0 & 0 & -2 \\ & 0 & -2 \\ & & 0 \end{pmatrix}$$

$$E_3^{(4)} : \begin{pmatrix} 1 & 1 & 1 & -1 \\ & 1 & 1 & -1 \\ & & 1 & -1 \\ & & & 0 \end{pmatrix} \quad E_4^{(4)} : \begin{pmatrix} 0 & 0 & 0 & 0 & 0 \\ & 0 & 0 & 0 & 0 \\ & & 0 & 0 & 0 \\ & & & 1 & -1 \\ & & & & 1 \end{pmatrix}.$$

(5th **blow-up**). In this case we blow-up a line $L \subset E_4^{(4)}$.

We have $\pi_5^*(E_i^{(4)}) = E_i^{(5)}$ for $i = 0, 1, 2, 3$, and $\pi_4^*(E_4^{(4)}) = E_4^{(5)} + E_5^{(5)}$. Then we have

$$E_0^{(5)} : \begin{pmatrix} -2 \end{pmatrix} \quad E_1^{(5)} : \begin{pmatrix} -1 & -3 \\ & -1 \end{pmatrix} \quad E_2^{(5)} : \begin{pmatrix} 0 & 0 & -2 \\ & 0 & -2 \\ & & 0 \end{pmatrix}$$

$$E_3^{(5)} : \begin{pmatrix} 1 & 1 & 1 & -1 \\ & 1 & 1 & -1 \\ & & 1 & -1 \\ & & & 0 \end{pmatrix} \quad E_4^{(5)} : \begin{pmatrix} 0 & 0 & 0 & 0 & 0 \\ & 0 & 0 & 0 & 0 \\ & & 0 & 0 & 0 \\ & & & 1 & -2 \\ & & & & 4 \end{pmatrix}$$

$$E_5^{(5)} : \begin{pmatrix} 0 & 0 & 0 & 0 & 0 & 0 \\ & 0 & 0 & 0 & 0 & 0 \\ & & 0 & 0 & 0 & 0 \\ & & & 0 & 1 & -1 \\ & & & & -2 & 1 \\ & & & & & 0 \end{pmatrix}.$$

The computations are made with the same techniques as in the proof of Lemma 4.3.14, using Proposition 1.2.24 for π_5, Proposition 1.2.25 for $\pi_4 \circ \pi_5$ (see Remark 1.2.26). We also use Proposition 1.2.27 and Proposition 1.2.2 to prove

$$E_3^{(5)} \cdot E_3^{(5)} \cdot E_5^{(5)} = E_3^{(5)}\Big|_{E_5^{(5)}} \cdot E_3^{(5)}\Big|_{E_5^{(5)}} = H \cdot H = 0,$$

$$E_4^{(5)} \cdot E_5^{(5)} \cdot E_5^{(5)} = E_5^{(5)}\Big|_{E_4^{(5)}} \cdot E_5^{(5)}\Big|_{E_4^{(5)}} = L \cdot L = 1,$$

where H is the divisor $\pi_5^{-1}(p)$ with p a generic point in L. The last equalities hold since H is a fiber of a ruled surface, and hence has null self-intersection, and L is a line in $\mathbb{P}^2$.

(6th **blow-up**). In this case we blow-up a smooth curve $C \subset E_5^{(5)}$.

We have $\pi_6^*(E_i^{(5)}) = E_i^{(6)}$ for $i = 0, 1, 2, 3, 4$, and $\pi_4^*(E_5^{(5)}) = E_5^{(6)} +$

$E_6^{(6)}$. Then we have

$$E_0^{(6)} : \begin{pmatrix} -2 \end{pmatrix} \qquad E_1^{(6)} : \begin{pmatrix} -1 & -3 \\ & -1 \end{pmatrix}$$

$$E_2^{(6)} : \begin{pmatrix} 0 & 0 & -2 \\ & 0 & -2 \\ & & 0 \end{pmatrix} \qquad E_3^{(6)} : \begin{pmatrix} 1 & 1 & 1 & -1 \\ & 1 & 1 & -1 \\ & & 1 & -1 \\ & & & 0 \end{pmatrix}$$

$$E_4^{(6)} : \begin{pmatrix} 0 & 0 & 0 & 0 & 0 \\ & 0 & 0 & 0 & 0 \\ & & 0 & 0 & 0 \\ & & & 1 & -2 \\ & & & & 4 \end{pmatrix} \qquad E_5^{(6)} : \begin{pmatrix} 0 & 0 & 0 & 0 & 0 & 0 \\ & 0 & 0 & 0 & 0 & 0 \\ & & 0 & 0 & 0 & 0 \\ & & & 0 & 1 & -2 \\ & & & & -2 & 1 \\ & & & & & 4 \end{pmatrix}$$

$$E_6^{(6)} : \begin{pmatrix} 0 & 0 & 0 & 0 & 0 & 0 & 0 \\ & 0 & 0 & 0 & 0 & 0 & 0 \\ & & 0 & 0 & 0 & 0 & 0 \\ & & & 0 & 0 & 1 & -1 \\ & & & & 0 & 0 & 0 \\ & & & & & -3 & 2 \\ & & & & & & -1 \end{pmatrix}.$$

As before, the computations are made with the same techniques as in the proof of Lemma 4.3.14, using Proposition 1.2.24 for π_6, Proposition 1.2.25 for $\pi_4 \circ \pi_5 \circ \pi_6$ (see Remark 1.2.26). We also use Proposition 1.2.27 and Proposition 1.2.2 to prove

$$E_3^{(6)} \cdot E_3^{(6)} \cdot E_6^{(6)} = E_3^{(6)}\Big|_{E_6^{(6)}} \cdot E_3^{(6)}\Big|_{E_6^{(6)}} = H \cdot H = 0,$$

$$E_5^{(6)} \cdot E_6^{(6)} \cdot E_6^{(6)} = E_6^{(6)}\Big|_{E_5^{(6)}} \cdot E_6^{(6)}\Big|_{E_5^{(6)}} = C \cdot C = 2,$$

where H is the divisor $\pi_6^{-1}(p)$ with p a generic point in C. For the last equalities, the first one follows again because H is a fiber of a ruled surface, while $C \cdot C = 2$ since $C = L_{-c}$, see Proposition 4.3.2(6th blow-up) and Proposition 1.4.26.

(**Gluing**) After the identification, we have that $E_0^{(6)}$ and $E_6^{(6)}$ glue together, while the other divisors are not affected. Then first of all

$$E_i \cdot E_j \cdot E_k = E_i^{(6)} \cdot E_j^{(6)} \cdot E_k^{(6)}$$

for every $i, j, k = 1, \ldots, 5$.

For other intersection numbers, we can look at the situation in the universal covering $\widetilde{X}$. If we consider a triple (E_i, E_j, E_k) with $k = 1, \ldots, 5$, then applying Proposition 1.2.27, we get

$$E_i \cdot E_j \cdot E_k = E_i|_{E_k} \cdot E_j\big|_{E_k}.$$

It follows that these intersection numbers depend only on the geometry in a neighborhood of E_k, and hence

$$E_0 \cdot E_j \cdot E_k = \left(E_0^{(6)} + E_6^{(6)}\right) \cdot E_j^{(6)} \cdot E_k^{(6)},$$
$$E_0 \cdot E_0 \cdot E_k = \left(E_0^{(6)} + E_6^{(6)}\right) \cdot \left(E_0^{(6)} + E_6^{(6)}\right) \cdot E_k^{(6)},$$

for every j and for $k = 1, \ldots 5$.

The last intersection number $E_0 \cdot E_0 \cdot E_0$ is obtained in the following way. To obtain E_0, we start from $E_6^{(6)}$, whose self-intersection number is $E_6^{(6)} \cdot E_6^{(6)} \cdot E_6^{(6)} = -1$ and we blow-up three points one over another, always in the strict transform of $E_6^{(6)}$, obtaining the divisor $E_6^{(9)}$ in the universal covering $\widetilde{X}$ of X. Since the projection $\omega : \widetilde{X} \to X$ gives a biholomorphism between a neighborhood of $E_6^{(9)}$ and a neighborhood of E_0, we get that their self-intersections are equal. Then thanks to Lemma 4.3.14, we have that

$$E_0 \cdot E_0 \cdot E_0 = E_6^{(9)} \cdot E_6^{(9)} \cdot E_6^{(9)} = E_6^{(6)} \cdot E_6^{(6)} \cdot E_6^{(6)} - 3 = -1 - 3 = -4.$$

□

4.3.4. Canonical bundle and Kodaira dimension

In this subsection we shall compute the canonical bundle of X, and show that the Kodaira dimension kod (X) of X is equal to $-\infty$.

Proposition 4.3.16. *Let X_j be the manifolds, and $E_i^{(j)}$ the divisors described in Definition 4.2.4.*

We have

(i) $\mathcal{K}_{X_0} = \mathcal{K}_{\mathbb{P}^3} = -4E_0^{(0)}$.
(ii) $\mathcal{K}_{X_1} = -4E_0^{(1)} - 2E_1^{(1)}$.
(iii) $\mathcal{K}_{X_2} = -4E_0^{(2)} - 2E_1^{(2)} - 4E_2^{(2)}$.
(iv) $\mathcal{K}_{X_3} = -4E_0^{(3)} - 2E_1^{(3)} - 4E_2^{(3)} - 8E_3^{(3)}$.
(v) $\mathcal{K}_{X_4} = -4E_0^{(4)} - 2E_1^{(4)} - 4E_2^{(4)} - 8E_3^{(4)} - 6E_4^{(4)}$.

(vi) $\mathcal{K}_{X_5} = -4E_0^{(5)} - 2E_1^{(5)} - 4E_2^{(5)} - 8E_3^{(5)} - 6E_4^{(5)} - 5E_5^{(5)}$.
(vii) $\mathcal{K}_{X_6} = -4E_0^{(6)} - 2E_1^{(6)} - 4E_2^{(6)} - 8E_3^{(6)} - 6E_4^{(6)} - 5E_5^{(6)} - 4E_6^{(6)}$.

Proof. The first statement is given by Example 1.2.18.(i). All the others are obtained step-by-step using Proposition 1.2.19. □

Proposition 4.3.17. *Let X be the Kato variety and $E_0, \ldots, E_5$ the divisors described in Definition* 4.2.6. *Then the canonical bundle of X is*

$$\mathcal{K}_X = -4E_0 - 2E_1 - 4E_2 - 8E_3 - 6E_4 - 5E_5.$$

Proof. Let us denote by $\widetilde{X}$ the universal covering of X, by $\omega : \widetilde{X} \to X$ the canonical projection, and $\widetilde{g} : \widetilde{X} \to \widetilde{X}$ the automorphism of the universal covering, and A_i as defined in Subsection 4.2.3. For canonical bundles, we have

$$\mathcal{K}_{\widetilde{X}} = \omega^* \mathcal{K}_X,$$

as line bundles.

Let us denote by $\widetilde{E}_i$ for $i = 1, \ldots, 5$ the divisors in $\widetilde{X}$ given by $E_i^{(6)}$ in A_0, by $\widetilde{E}_0$ the divisor in $\widetilde{X}$ given by gluing $E_0^{(6)} \in A_0$ and $E_6^{(6)} \in A_{-1}$, and $\widetilde{E}_{i+6k} = g^{ok}(E_i)$ for $i = 0, \ldots, 5$ and $k \in \mathbb{Z}$.

With the same techniques as in Proposition 4.3.16, we find that the canonical bundle of $\widetilde{X}$ is given by

$$\mathcal{K}_{\widetilde{X}} = \sum_{i \in \mathbb{Z}} a_i \widetilde{E}_i,$$

with

$$\begin{aligned} a_{6k} &= -4, & a_{6k+1} &= -2, & a_{6k+2} &= -4, \\ a_{6k+3} &= -8, & a_{6k+4} &= -6, & a_{6k+5} &= -5, \end{aligned}$$

for every $k \in \mathbb{Z}$.

In particular $\mathcal{K}_{\widetilde{X}} = \left[\widetilde{E}\right]$ for a suitable divisor $\widetilde{E}$ that is invariant for the action of g^*.

It follows that

$$\mathcal{K}_X = \left[g_* \widetilde{E}\right],$$

and we are done. □

Lemma 4.3.18. *Let X be a compact complex manifold, and E a line bundle over X. If $-E$ is effective and non-trivial, then $h^0(X, \mathcal{O}(E)) = 0$, i.e., E has no non-null holomorphic sections.*

Proof. By absurd, let us suppose that E admits a non-trivial global section s_+. Since $-E$ is effective, it admits a non-trivial global section s_-. We can then consider $s = s_+ \otimes s_-$ as a non-trivial global section of $E - E = 0$, *i.e.*, $s \not\equiv 0$ is a holomorphic function $s : X \to \mathbb{C}$. Being X compact, then s is a constant function, *e.g.*, $s \equiv 1$. If we consider a local chart U, then if $s_\pm$ is given by $s_\pm(x) = (x, h_\pm(x))$, then we have that

$$1 = s(x) = h_+(x) \cdot h_-(x).$$

It follows that s_+ (resp. s_-) is a holomorphic and everywhere non-zero section of E (resp., $-E$), that hence is the trivial line bundle, in contradiction with our assumptions. □

Corollary 4.3.19. *Let X be the Kato variety constructed in Subsection* 4.2.3. *Then* $\operatorname{kod}(X) = -\infty$.

Proof. Thanks to Proposition 4.3.17, $-\mathcal{K}_X$, and hence $-m\mathcal{K}_X$ for every $m \in \mathbb{N}^*$, is an effective and non-trivial line bundle. Then thanks to Lemma 4.3.18, $P_m(X) = 0$ for every $m \in \mathbb{N}^*$, and we are done. □

Remark 4.3.20. The Kodaira dimension is a birational invariant (see [39, pp. 131–133]), and hence every birationally equivalent model of X has the same Kodaira dimension $\operatorname{kod}() = -\infty$.

4.3.5. Canonical and normal bundles

Now we want to compute normal and canonical bundles for all the divisors we constructed before.

First of all, we shall underline the algebraic properties of all divisors, that shall be useful in the whole chapter.

Proposition 4.3.21. *Let us use the same notations as in Definition* 4.3.1. *Then we have the following properties.*

(i) *All effective divisors in $\mathbb{P}^2$ are positive multiples of $D_0^{(0)}$, whose self-intersection is $D_0^{(0)} \cdot D_0^{(0)} = 1$ Moreover its canonical bundle is given by*

$$\mathcal{K}_{\mathbb{P}^2} = -3D_0^{(0)}.$$

(ii) *All effective divisors in $\mathbb{P}^2_1$ are a positive linear combination of $D_0^{(1)}$ and $D_1^{(1)}$. The intersection numbers are given by the matrix*

$$\begin{pmatrix} 0 & 1 \\ & -1 \end{pmatrix}.$$

Moreover its canonical bundle is given by

$$\mathcal{K}_{\mathbb{P}^2_1} = -3D_0^{(1)} - 2D_1^{(1)}.$$

(iii) *All effective divisors in* $\mathbb{P}^2_2$ *are a positive linear combination of* $D_0^{(2)}, D_1^{(2)}$ *and* $D_2^{(2)}$. *The intersection numbers are given by the matrix*

$$\begin{pmatrix} -1 & 0 & 1 \\ & -2 & 1 \\ & & -1 \end{pmatrix}.$$

Moreover its canonical bundle is given by

$$\mathcal{K}_{\mathbb{P}^2_2} = -3D_0^{(2)} - 2D_1^{(2)} - 4D_2^{(2)}.$$

(iv) *All effective divisors in* $\mathbb{P}^2_{3a}$ *are a positive linear combination of* $D_i^{(3a)}$ *for* $i = 0, \dots, 3$. *The intersection numbers are given by the matrix*

$$\begin{pmatrix} -1 & 0 & 1 & 0 \\ & -3 & 0 & 1 \\ & & -2 & 1 \\ & & & -1 \end{pmatrix}.$$

Moreover its canonical bundle is given by

$$\mathcal{K}_{\mathbb{P}^2_{3a}} = -3D_0^{(3a)} - 2D_1^{(3a)} - 4D_2^{(3a)} - 5D_3^{(3a)}.$$

(v) *All effective divisors in* $\mathbb{P}^2_{3b}$ *are a positive linear combination of* $D_i^{(3b)}$ *for* $i = 0, \dots, 3$. *The intersection numbers are given by the matrix*

$$\begin{pmatrix} -2 & 0 & 0 & 1 \\ & -2 & 1 & 0 \\ & & -2 & 1 \\ & & & -1 \end{pmatrix}.$$

Moreover its canonical bundle is given by

$$\mathcal{K}_{\mathbb{P}^2_{3b}} = -3D_0^{(3b)} - 2D_1^{(3b)} - 4D_2^{(3b)} - 6D_3^{(3b)}.$$

(vi) *For every* $n \in \mathbb{N}$, *all effective divisors in* $\mathbb{F}_n$ *are a positive linear combination of* L_∞ *and* H. *The intersection numbers are given by the matrix*

$$\begin{pmatrix} -n & 1 \\ & 0 \end{pmatrix}.$$

Moreover its canonical bundle is given by

$$\mathcal{K}_{\mathbb{F}_n} = -2[L_\infty] - (2+n)[H].$$

(vii) *If $n \geq 1$, all effective divisors in $(\mathbb{F}_n)_{3a}$ are a positive linear combination of L_∞, $H^{(3a)}$ and $D_i^{(3a)}$ for $i = 1, 2, 3$. The intersection numbers are given by the matrix*

$$\begin{pmatrix} -n & 1 & 0 & 0 & 0 \\ & -2 & 0 & 1 & 0 \\ & & -3 & 0 & 1 \\ & & & -2 & 1 \\ & & & & -1 \end{pmatrix}.$$

Moreover its canonical bundle is given by

$$\begin{aligned} \mathcal{K}_{(\mathbb{F}_n)_{3a}} = & -2[L_\infty] - (2+n)\left[H^{(3a)}\right] \\ & - (1+n)\left[D_1^{(3a)}\right] - (2+2n)\left[D_2^{(3a)}\right] - (2+3n)\left[D_3^{(3a)}\right]. \end{aligned}$$

Proof. All the computations on intersection numbers are made by using Corollary 1.2.28, and Proposition 1.4.26 for ruled surfaces.

All the computations on canonical bundles are made by using Proposition 1.2.19, Example 1.2.18 and Proposition 1.4.28 for ruled surfaces.

All the considerations on effective divisors are trivial, but for the last one. Indeed, the point in H that we blow-up to obtain $(\mathbb{F}_n)_1$ in Definition 4.3.1 belongs to L_c for a suitable $c \in \mathbb{C}$. Then, we only have to check that the strict transform $L_c^{(1)}$ of L_c can be obtained by a positive combination of L_∞, $H^{(1)}$ and $D_1^{(1)}$. But we have

$$L_c^{(1)} + D_1^{(1)} = L_c = L_\infty + nH = L_\infty + nH^{(1)} + nD_1^{(1)},$$

hence $L_c^{(1)} = L_\infty + nH^{(1)} + (n-1)D_1^{(1)}$, and we are done since $n \geq 1$.

Since the point in $(\mathbb{F}_n)_1$ we blow-up to get $(\mathbb{F}_n)_2$ does not belong to $L_c^{(1)}$, we are done. □

Proposition 4.3.22. *Let X be the Kato variety and $E_0, \ldots, E_5$ the divisors described in Definition 4.2.6. Moreover, let us use notations as in Definition 4.3.1. Then for the normal bundles $\mathcal{N}_{E_i}$ we have*

$$\mathcal{N}_{E_1} = -\left[D_0^{(2)}\right] - 2\left[D_1^{(2)}\right] - 4\left[D_2^{(2)}\right],$$

$$\mathcal{N}_{E_2} = -\left[D_0^{(1)}\right] - 2\left[D_1^{(1)}\right],$$

$$\mathcal{N}_{E_3} = -\left[D_0^{(3b)}\right] - 2\left[D_1^{(3b)}\right] - 3\left[D_2^{(3b)}\right] - 4\left[D_3^{(3b)}\right],$$

$$\mathcal{N}_{E_4} = -2\left[D_0^{(0)}\right] = \mathcal{O}(-2),$$

$$\mathcal{N}_{E_5} = -2[L_\infty] - 3[H],$$

$$\mathcal{N}_{E_0} = -[L_\infty] - \left[H^{(3a)}\right] - 2\left[D_1^{(3a)}\right] - 4\left[D_2^{(3a)}\right] - 7\left[D_3^{(3a)}\right].$$

Proof. We proceed step-by-step, as usual, computing all normal bundles for $E_i^{(j)}$ for every $0 \leq i \leq j \leq 6$, and we shall finally deal with the gluing. We shall heavily use the Adjunction Formulae given by Propositions 1.2.20 and 1.2.21, and we shall use the results in Proposition 4.3.16 and Proposition 4.3.21.

We start from $E_0^{(0)}$. Using Propositions 1.2.21 and 1.2.20, we get

$$\begin{array}{l} \mathcal{K}_{E_0^{(0)}} = \left(\mathcal{K}_{X_0} \otimes \left[E_0^{(0)}\right]\right)\Big|_{E_0^{(0)}} = -3\left[E_0^{(0)}\right]\Big|_{E_0^{(0)}} = -3\mathcal{N}_{E_0^{(0)}} \\ \quad || \\ \mathcal{K}_{\mathbb{P}^2} = -3\left[D_0^{(0)}\right], \end{array}$$

and hence $\mathcal{N}_{E_0^{(0)}} = \left[D_0^{(0)}\right] = \mathcal{O}(1)$.

(1st **blow-up**). Thanks to Proposition 4.3.21.(i), we get

$$\begin{array}{l} \mathcal{K}_{E_1^{(1)}} = \left(\mathcal{K}_{X_1} \otimes \left[E_1^{(1)}\right]\right)\Big|_{E_1^{(1)}} = -4\left[E_0^{(1)}\right]\Big|_{E_1^{(1)}} - \mathcal{N}_{E_1^{(1)}} \\ \quad || \\ \mathcal{K}_{\mathbb{P}^2} = -3\left[D_0^{(0)}\right]. \end{array}$$

In this case, $\left[E_0^{(1)}\right]\Big|_{E_1^{(1)}} = \left[D_0^{(0)}\right]$ (see Proposition 1.2.2), and hence $\mathcal{N}_{E_1^{(1)}} = -\left[D_0^{(0)}\right] = \mathcal{O}(-1)$.

In particular, we got the following statement.

Lemma 4.3.23. *Let $\pi : \widetilde{X} \to X$ the blow-up of a 3-manifold X along a point p, and $E \cong \mathbb{P}^2$ its exceptional divisor. Then the normal bundle of E in $\widetilde{X}$ is*

$$\mathcal{N}_E = -\left[D_0^{(0)}\right] = \mathcal{O}(-1).$$

Again, using Proposition 4.3.21.(ii), we get

$$\begin{array}{l} \mathcal{K}_{E_0^{(1)}} = \left(\mathcal{K}_{X_1} \otimes \left[E_0^{(1)}\right]\right)\Big|_{E_0^{(1)}} = -3\mathcal{N}_{E_0^{(1)}} - 2\left[E_1^{(1)}\right]\Big|_{E_0^{(1)}} \\ \quad || \\ \mathcal{K}_{\mathbb{P}_1^2} = -3\left[D_0^{(1)}\right] - 2\left[D_1^{(1)}\right]. \end{array}$$

In this case, $\left[E_1^{(1)}\right]\Big|_{E_0^{(1)}} = \left[D_1^{(1)}\right]$, and hence $\mathcal{N}_{E_0^{(1)}} = \left[D_0^{(1)}\right]$.

(2nd **blow-up**). First of all, thanks to Lemma 4.3.23, we get $\mathcal{N}_{E_2^{(2)}} = -\left[D_0^{(0)}\right] = \mathcal{O}(-1)$.

Using Proposition 4.3.21.(ii), we get

$$\begin{aligned} \mathcal{K}_{E_1^{(2)}} &= \left(\mathcal{K}_{X_2} \otimes \left[E_1^{(2)}\right]\right)\Big|_{E_1^{(2)}} = -4\left[E_0^{(2)}\right]\Big|_{E_1^{(2)}} - \mathcal{N}_{E_1^{(2)}} - 4\left[E_2^{(2)}\right]\Big|_{E_1^{(2)}} \\ &\,|| \\ \mathcal{K}_{\mathbb{P}_1^2} &= -3\left[D_0^{(1)}\right] - 2\left[D_1^{(1)}\right]. \end{aligned}$$

In this case, $\left[E_0^{(2)}\right]\Big|_{E_1^{(2)}} = \left[D_0^{(0)}\right]$ and $\left[E_2^{(2)}\right]\Big|_{E_1^{(2)}} = \left[D_1^{(1)}\right]$, and hence $\mathcal{N}_{E_1^{(2)}} = -\left[D_0^{(1)}\right] - 2\left[D_1^{(1)}\right]$.

In particular, we got the following statement.

Lemma 4.3.24. *Let $\pi = \pi_1 \circ \pi_2 : \widetilde{X} \to X$ the modification of a 3-manifold X given by blowing-up a point $p_1 \in X$, and getting an exceptional divisor E_1, and then another point $p_2 \in E_1$, getting an exceptional divisor E_2. Let $\widetilde{E}_1 \cong \mathbb{P}_1^2$ be the strict transform of E_1 through π_2. Then the normal bundle of $\widetilde{E}_1$ in $\widetilde{X}$ is*

$$\mathcal{N}_{\widetilde{E}_1} = -\left[D_0^{(1)}\right] - 2\left[D_1^{(1)}\right].$$

Using Proposition 4.3.21.(ii), we get

$$\begin{aligned} \mathcal{K}_{E_0^{(2)}} &= \left(\mathcal{K}_{X_2} \otimes \left[E_0^{(2)}\right]\right)\Big|_{E_0^{(2)}} = -3\mathcal{N}_{E_0^{(2)}} - 2\left[E_1^{(2)}\right]\Big|_{E_0^{(2)}} - 4\left[E_2^{(2)}\right]\Big|_{E_0^{(2)}} \\ &\,|| \\ \mathcal{K}_{\mathbb{P}_2^2} &= -3\left[D_0^{(2)}\right] - 2\left[D_1^{(2)}\right] - 4\left[D_2^{(2)}\right]. \end{aligned}$$

In this case, $\left[E_i^{(2)}\right]\Big|_{E_0^{(2)}} = \left[D_i^{(2)}\right]$ for $i = 1, 2$, and hence $\mathcal{N}_{E_0^{(2)}} = \left[D_0^{(2)}\right]$.

(3rd **blow-up**). First of all, thanks to Lemmas 4.3.23 and 4.3.24 respectively, we get $\mathcal{N}_{E_3^{(3)}} = -\left[D_0^{(0)}\right] = \mathcal{O}(-1)$ and $\mathcal{N}_{E_2^{(3)}} = -\left[D_0^{(1)}\right] - 2\left[D_1^{(1)}\right]$.

Using Proposition 4.3.21.(iii), we get

$$\begin{aligned}\mathcal{K}_{E_1^{(3)}} &= \left(\mathcal{K}_{X_3} \otimes \left[E_1^{(3)}\right]\right)\Big|_{E_1^{(3)}} \\ &= -4\left[E_0^{(3)}\right]\Big|_{E_1^{(3)}} - \mathcal{N}_{E_1^{(3)}} - 4\left[E_2^{(3)}\right]\Big|_{E_1^{(3)}} - 8\left[E_3^{(3)}\right]\Big|_{E_1^{(3)}}\end{aligned}$$

$$||$$

$$\mathcal{K}_{\mathbb{P}^2_2} = -3\left[D_0^{(2)}\right] - 2\left[D_1^{(2)}\right] - 4\left[D_2^{(2)}\right].$$

In this case, $\left[E_0^{(3)}\right]\Big|_{E_1^{(3)}} = \left[D_0^{(2)}\right]$, and $\left[E_i^{(3)}\right]\Big|_{E_1^{(3)}} = \left[D_{i-1}^{(2)}\right]$ for $i = 2, 3$, and hence $\mathcal{N}_{E_1^{(3)}} = -\left[D_0^{(2)}\right] - 2\left[D_1^{(2)}\right] - 4\left[D_2^{(2)}\right]$.

Using Proposition 4.3.21.(iv), we get

$$\begin{aligned}\mathcal{K}_{E_0^{(3)}} &= \left(\mathcal{K}_{X_3} \otimes \left[E_0^{(3)}\right]\right)\Big|_{E_0^{(3)}} \\ &= -3\mathcal{N}_{E_0^{(3)}} - 2\left[E_1^{(3)}\right]\Big|_{E_0^{(3)}} - 4\left[E_2^{(3)}\right]\Big|_{E_0^{(3)}} - 8\left[E_3^{(3)}\right]\Big|_{E_0^{(3)}}\end{aligned}$$

$$||$$

$$\mathcal{K}_{\mathbb{P}^2_{3a}} = -3\left[D_0^{(3a)}\right] - 2\left[D_1^{(3a)}\right] - 4\left[D_2^{(3a)}\right] - 5\left[D_3^{(3a)}\right].$$

In this case $\left[E_i^{(3)}\right]\Big|_{E_0^{(3)}} = \left[D_i^{(3a)}\right]$ for $i = 1, 2, 3$, and hence $\mathcal{N}_{E_0^{(3)}} = \left[D_0^{(3a)}\right] - \left[D_3^{(3a)}\right]$.

(4th **blow-up**). Thanks to Lemmas 4.3.23 and 4.3.24 respectively, we get $\mathcal{N}_{E_4^{(4)}} = -\left[D_0^{(0)}\right] = \mathcal{O}(-1)$ and $\mathcal{N}_{E_3^{(4)}} = -\left[D_0^{(1)}\right] - 2\left[D_1^{(1)}\right]$.

For the other bundles, since we are blowing-up a point that does not belong to $E_i^{(3)}$ for $i = 0, 1, 2$, then $\mathcal{N}_{E_i^{(4)}} = \mathcal{N}_{E_i^{(3)}}$ again for $i = 0, 1, 2$.

(5th **blow-up**). We are blowing up a line $L \subset E_4^{(4)}$ this time, so we have to make some computations on this line before computing normal bundles for all the divisors.

Since $L \cong \mathbb{P}^1$ is a line in $E_4^{(4)}$, then $\mathcal{N}_{L\subset E_4^{(4)}} = [p] = \mathcal{O}(1)$.

We could prove this also by using Propositions 1.2.20 and 1.2.21 (Adjunction formulae):

$$\mathcal{K}_{L\subset E_4^{(4)}} = \left(\mathcal{K}_{E_4^{(4)}} \otimes [L]\right)\Big|_L = -3\left[D_0^{(0)}\right]\Big|_L + \mathcal{N}_{L\subset E_4^{(4)}} = -2\mathcal{N}_{L\subset E_4^{(4)}}$$

$$||$$

$$\mathcal{K}_{\mathbb{P}^1} = -2[p],$$

and again $\mathcal{N}_{L\subset E_4^{(4)}} = [p] = \mathcal{O}\,(1)$.

Moreover, we have

$$\mathcal{N}_{E_4^{(4)}\subset X_4}\Big|_L = -\Big[D_0^{(0)}\Big]\Big|_L = -[p] = \mathcal{O}(-1).$$

Since we have

$$\mathcal{N}_{L\subset X_4} = \mathcal{N}_{L\subset E_4^{(4)}} \oplus \mathcal{N}_{E_4^{(4)}\subset X_4}\Big|_L = \mathcal{O}(1)\oplus\mathcal{O}(-1),$$

we get (see Definition 1.4.23 and Corollary 1.4.25) that $E_5^{(5)} \cong \mathbb{F}_{1,-1} \cong \mathbb{F}_2$, and $E_4^{(5)}\Big|_{E_5^{(5)}} = L_\infty$.

Now we can compute canonical and normal bundles for $E_5^{(5)}$. Using also Proposition 4.3.21.(vi), we get

$$\begin{aligned}
\mathcal{K}_{E_5^{(5)}} &= \Big(\mathcal{K}_{X_5}\otimes\Big[E_5^{(5)}\Big]\Big)\Big|_{E_5^{(5)}}\\
&= -4\Big[E_0^{(5)}\Big]\Big|_{E_5^{(5)}} - 2\Big[E_1^{(5)}\Big]\Big|_{E_5^{(5)}} - 4\Big[E_2^{(5)}\Big]\Big|_{E_5^{(5)}}\\
&\quad - 8\Big[E_3^{(5)}\Big]\Big|_{E_5^{(5)}} - 6\Big[E_4^{(5)}\Big]\Big|_{E_5^{(5)}} - 4\mathcal{N}_{E_5^{(5)}}\\
&\,\|\\
\mathcal{K}_{\mathbb{F}_2} &= -2[L_\infty] - 4[H].
\end{aligned}$$

Then we have $\Big[E_i^{(5)}\Big]\Big|_{E_5^{(5)}} = 0$ for $i = 0, 1, 2$, $\Big[E_3^{(5)}\Big]\Big|_{E_5^{(5)}} = [H]$ and $\Big[E_4^{(5)}\Big]\Big|_{E_5^{(5)}} = [L_\infty]$; hence $\mathcal{N}_{E_5^{(5)}} = -[L_\infty] - [H]$.

Using Proposition 4.3.21.(i), we get

$$\begin{aligned}
\mathcal{K}_{E_4^{(5)}} &= \Big(\mathcal{K}_{X_5}\otimes\Big[E_4^{(5)}\Big]\Big)\Big|_{E_4^{(5)}}\\
&= -4\Big[E_0^{(5)}\Big]\Big|_{E_4^{(5)}} - 2\Big[E_1^{(5)}\Big]\Big|_{E_4^{(5)}} - 4\Big[E_2^{(5)}\Big]\Big|_{E_4^{(5)}}\\
&\quad - 8\Big[E_3^{(5)}\Big]\Big|_{E_4^{(5)}} - 5\mathcal{N}_{E_4^{(5)}} - 5\Big[E_5^{(5)}\Big]\Big|_{E_4^{(5)}}\\
&\,\|\\
\mathcal{K}_{\mathbb{P}^2} &= -3\Big[D_0^{(0)}\Big],
\end{aligned}$$

Then we have $\Big[E_i^{(5)}\Big]\Big|_{E_4^{(5)}} = 0$ for $i = 0, 1, 2$, $\Big[E_i^{(5)}\Big]\Big|_{E_4^{(5)}} = \Big[D_0^{(0)}\Big]$ for $i = 3, 5$, and hence $\mathcal{N}_{E_4^{(5)}} = -2\Big[D_0^{(0)}\Big] = \mathcal{O}(-2)$.

Using Proposition 4.3.21.(iii), we get

$$\begin{aligned}
\mathcal{K}_{E_3^{(5)}} &= \left(\mathcal{K}_{X_5} \otimes \left[E_3^{(5)}\right]\right)\Big|_{E_3^{(5)}} \\
&= -4\left[E_0^{(5)}\right]\Big|_{E_3^{(5)}} - 2\left[E_1^{(5)}\right]\Big|_{E_3^{(5)}} - 4\left[E_2^{(5)}\right]\Big|_{E_3^{(5)}} \\
&\quad - 7\mathcal{N}_{E_3^{(5)}} - 6\left[E_4^{(5)}\right]\Big|_{E_3^{(5)}} - 5\left[E_5^{(5)}\right]\Big|_{E_3^{(5)}}
\end{aligned}$$

$$||$$

$$\mathcal{K}_{\mathbb{P}^2_2} = -3\left[D_0^{(2)}\right] - 2\left[D_1^{(2)}\right] - 4\left[D_2^{(2)}\right].$$

Then we have $\left[E_i^{(5)}\right]\Big|_{E_3^{(5)}} = \left[D_0^{(0)}\right] = \left[D_0^{(2)} + D_1^{(2)} + 2D_2^{(2)}\right]$ for $i = 0, 1, 2$, while $\left[E_4^{(5)}\right]\Big|_{E_3^{(5)}} = \left[D_1^{(2)}\right]$ and $\left[E_5^{(5)}\right]\Big|_{E_3^{(5)}} = \left[D_2^{(2)}\right]$; hence $\mathcal{N}_{E_3^{(5)}} = -\left[D_0^{(2)}\right] - 2\left[D_1^{(2)}\right] - 3\left[D_2^{(2)}\right]$.

As in the previous step, the fifth blow-up does not change the normal bundles of the first three divisors, and hence $\mathcal{N}_{E_i^{(5)}} = \mathcal{N}_{E_i^{(4)}}$ again for $i = 0, 1, 2$.

(6th **blow-up**). We are blowing-up the curve $C = L_{-c} \subset E_5^{(5)} \cong \mathbb{F}_2$ (see also Proposition 4.3.2(6th blow-up)). The curve C does not intersect L_∞, and intersect transversely H in one point.

Using Propositions 1.2.20 and 1.2.21 (Adjunction formulae) we get

$$\begin{aligned}
\mathcal{K}_{C\subset E_5^{(5)}} &= \left(\mathcal{K}_{E_5^{(5)}} \otimes [C]\right)\Big|_C = -4[H]|_C - 2[L_\infty]|_C + \mathcal{N}_{C\subset E_5^{(5)}} \\
&= -4[p] + \mathcal{N}_{C\subset E_5^{(5)}}
\end{aligned}$$

$$||$$

$$\mathcal{K}_{\mathbb{P}^1} = -2[p],$$

and hence $\mathcal{N}_{C\subset E_5^{(5)}} = 2[p] = \mathcal{O}(2)$ (this result follows also from Proposition 1.4.26). Moreover, we have

$$\mathcal{N}_{E_5^{(5)}\subset X_5}\Big|_C = -[L_\infty + H]|_C = -[p] = \mathcal{O}(-1).$$

Again we have

$$\mathcal{N}_{C\subset X_5} = \mathcal{N}_{C\subset E_5^{(5)}} \oplus \mathcal{N}_{E_5^{(5)}\subset X_5}\Big|_C = \mathcal{O}(2) \oplus \mathcal{O}(-1),$$

so (see again Definition 1.4.23 and Corollary 1.4.25) we have that $E_6^{(6)} \cong \mathbb{F}_{2,-1} \cong \mathbb{F}_3$ and $E_5^{(6)}\Big|_{E_6^{(6)}} = L_\infty$.

Now we can compute all canonical and normal bundles. This blow-up does not modify the divisors $E_i^{(5)}$ for $i = 0, 1, 2, 4$, and so neither the normal bundles associated to them.

From Proposition 4.3.21.(vi) we get

$$\begin{aligned}
\mathcal{K}_{E_6^{(6)}} &= \left(\mathcal{K}_{X_6} \otimes \left[E_6^{(6)}\right]\right)\Big|_{E_6^{(6)}} \\
&= -4\left[E_0^{(6)}\right]\Big|_{E_6^{(6)}} - 2\left[E_1^{(6)}\right]\Big|_{E_6^{(6)}} - 4\left[E_2^{(6)}\right]\Big|_{E_6^{(6)}} \\
&\quad - 8\left[E_3^{(6)}\right]\Big|_{E_6^{(6)}} - 6\left[E_4^{(6)}\right]\Big|_{E_6^{(6)}} - 5\left[E_5^{(6)}\right]\Big|_{E_6^{(6)}} - 3\mathcal{N}_{E_6^{(6)}} \\
\| \\
\mathcal{K}_{\mathbb{F}_3} &= -2[L_\infty] - 5[H].
\end{aligned}$$

Then we have $\left[E_i^{(6)}\right]\Big|_{E_6^{(6)}} = 0$ for $i = 0, 1, 2, 4$, $\left[E_3^{(6)}\right]\Big|_{E_6^{(6)}} = [H]$ and $\left[E_5^{(6)}\right]\Big|_{E_6^{(6)}} = [L_\infty]$; hence $\mathcal{N}_{E_6^{(6)}} = -[L_\infty] - [H]$.

Again also thanks to Proposition 4.3.21.(vi) we get

$$\begin{aligned}
\mathcal{K}_{E_5^{(6)}} &= \left(\mathcal{K}_{X_6} \otimes \left[E_5^{(6)}\right]\right)\Big|_{E_5^{(6)}} \\
&= -4\left[E_0^{(6)}\right]\Big|_{E_5^{(6)}} - 2\left[E_1^{(6)}\right]\Big|_{E_5^{(6)}} - 4\left[E_2^{(6)}\right]\Big|_{E_5^{(6)}} \\
&\quad - 8\left[E_3^{(6)}\right]\Big|_{E_5^{(6)}} - 6\left[E_4^{(6)}\right]\Big|_{E_5^{(6)}} - 4\mathcal{N}_{E_5^{(6)}} - 4\left[E_6^{(6)}\right]\Big|_{E_5^{(6)}} \\
\| \\
\mathcal{K}_{\mathbb{F}_2} &= -2[L_\infty] - 4[H].
\end{aligned}$$

Then we have $\left[E_i^{(6)}\right]\Big|_{E_5^{(6)}} = 0$ for $i = 0, 1, 2$, $\left[E_3^{(6)}\right]\Big|_{E_5^{(6)}} = [H]$, $\left[E_4^{(6)}\right]\Big|_{E_5^{(6)}} = [L_\infty]$ and $\left[E_6^{(6)}\right]\Big|_{E_5^{(6)}} = [C = L_{-c}] = [L_\infty + 2H]$, thanks to Proposition 1.4.26. Hence $\mathcal{N}_{E_5^{(6)}} = -2[L_\infty] - 3[H]$.

Finally, thanks to Proposition 4.3.21.(v), we get

$$\begin{aligned}
\mathcal{K}_{E_3^{(6)}} &= \left(\mathcal{K}_{X_6} \otimes \left[E_3^{(6)}\right]\right)\Big|_{E_3^{(6)}} \\
&= -4\left[E_0^{(6)}\right]\Big|_{E_3^{(6)}} - 2\left[E_1^{(6)}\right]\Big|_{E_3^{(6)}} - 4\left[E_2^{(6)}\right]\Big|_{E_3^{(6)}} - 7\mathcal{N}_{E_3^{(6)}} \\
&\quad - 6\left[E_4^{(6)}\right]\Big|_{E_3^{(6)}} - 5\left[E_5^{(6)}\right]\Big|_{E_3^{(6)}} - 4\left[E_6^{(6)}\right]\Big|_{E_3^{(6)}} \\
\| \\
\mathcal{K}_{\mathbb{P}^2_{3b}} &= -3\left[D_0^{(3b)}\right] - 2\left[D_1^{(3b)}\right] - 4\left[D_2^{(3b)}\right] - 6\left[D_3^{(3b)}\right].
\end{aligned}$$

Then we have $\left[E_i^{(6)}\right]\Big|_{E_3^{(6)}} = \left[D_0^{(0)}\right] = \left[D_0^{(3b)} + D_1^{(3b)} + 2D_2^{(3b)} + 3D_3^{(3b)}\right]$ for $i = 0, 1, 2$, $\left[E_i^{(6)}\right]\Big|_{E_3^{(6)}} = \left[D_{i-3}^{(3b)}\right]$ for $i = 4, 5, 6$; hence $\mathcal{N}_{E_3^{(6)}} = -\left[D_0^{(3b)}\right] - 2\left[D_1^{(3b)}\right] - 3\left[D_2^{(3b)}\right] - 4\left[D_3^{(3b)}\right]$.

(**Gluing**). The gluing does not change the normal bundles for the divisors $E_i^{(6)}$ for $i = 1, \ldots, 5$, so we just have to compute $\mathcal{N}_{E_0}$.

As we saw in Subsection 4.3.3, E_0 is obtained from $E_6^{(6)}$ by blowing-up three points one over another. So we can continue blowing up, and making computations as before. Let then X_9 and $E_i^{(9)}$ for $i = 0, \ldots, 9$ be defined as in Definition 4.2.9; we get

$$\begin{aligned}\mathcal{K}_{X_9} = {} & -4E_0^{(9)} - 2E_1^{(9)} - 4E_2^{(9)} - 8E_3^{(9)} - 6E_4^{(9)} - 5E_5^{(9)} \\ & - 4E_6^{(9)} - 2E_7^{(9)} - 4E_8^{(9)} - 8E_9^{(9)},\end{aligned}$$

using as usual Proposition 1.2.19, where as usual $E_i^{(9)}$ denotes the strict transform of $E_i^{(6)}$ for $i = 0, \ldots, 6$, and $E_i^{(9)}$ for $i = 7, 8, 9$ are the new exceptional divisors created by blowing up.

Using Proposition 4.3.21.(vii), we have

$$\begin{aligned}\mathcal{K}_{E_0} &= \left(\mathcal{K}_{X_9} \otimes \left[E_6^{(9)}\right]\right)\Big|_{E_6^{(9)}} \\ &= -4\left[E_0^{(9)}\right]\Big|_{E_6^{(9)}} - 2\left[E_1^{(9)}\right]\Big|_{E_6^{(9)}} - 4\left[E_2^{(9)}\right]\Big|_{E_6^{(9)}} \\ &\quad - 8\left[E_3^{(9)}\right]\Big|_{E_6^{(9)}} - 6\left[E_4^{(9)}\right]\Big|_{E_6^{(9)}} - 5\left[E_5^{(9)}\right]\Big|_{E_6^{(9)}} - 3\mathcal{N}_{E_6^{(9)}} \\ &\quad - 2\left[E_7^{(9)}\right]\Big|_{E_6^{(9)}} - 4\left[E_2^{(9)}\right]\Big|_{E_8^{(9)}} - 8\left[E_3^{(9)}\right]\Big|_{E_9^{(9)}} \\ & \Vert \\ \mathcal{K}_{(\mathbb{F}_3)_{3a}} &= -2[L_\infty] - 5\left[H^{(3a)}\right] - 4\left[D_1^{(3a)}\right] - 8\left[D_2^{(3a)}\right] - 11\left[D_3^{(3a)}\right].\end{aligned}$$

Now we have that $\left[E_i^{(9)}\right]\Big|_{E_6^{(9)}} = 0$ for $i = 0, 1, 2, 4$, while $\left[E_3^{(9)}\right]\Big|_{E_6^{(9)}} = [H] = \left[H^{(3a)} + D_1^{(3a)} + 2D_2^{(3a)} + 3D_3^{(3a)}\right]$, $\left[E_5^{(9)}\right]\Big|_{E_6^{(9)}} = [L_\infty]$, and $\left[E_{6+i}^{(9)}\right]\Big|_{E_6^{(9)}} = \left[D_i^{(3a)}\right]$ for $i = 1, 2, 3$. Making all these substitutions, we get $\mathcal{N}_{E_0} = -[C'] - \left[H^{(3a)}\right] - 2\left[D_1^{(3a)}\right] - 4\left[D_2^{(3a)}\right] - 7\left[D_3^{(3a)}\right]$. □

Remark 4.3.25. Corollary 1.2.29 gives a control on the computation of normal bundles and intersection numbers. Let us show some examples.

(i) For $E_1^{(2)} \cong \mathbb{P}_1^2$ we have $\mathcal{N}_{E_1^{(2)}} = -\left[D_0^{(1)}\right] - 2\left[D_1^{(1)}\right]$. Thanks to Proposition 4.3.21.(ii), we have

$$\begin{aligned}\left(\mathcal{N}_{E_1^{(2)}}\right)^2 &= \left(-D_0^{(1)} - 2D_1^{(1)}\right)^2 \\ &= \left(D_0^{(1)}\right)^2 + 4D_0^{(1)} \cdot D_1^{(1)} + 4\left(D_1^{(1)}\right)^2 \\ &= 1 \cdot 0 + 4 \cdot 1 + 4 \cdot (-1) \\ &= 0 = \left(E_1^{(2)}\right)^3.\end{aligned}$$

(ii) For $E_1^{(3)} \cong \mathbb{P}_2^2$ we have $\mathcal{N}_{E_1^{(3)}} = -\left[D_0^{(2)}\right] - 2\left[D_1^{(2)}\right] - 4\left[D_2^{(2)}\right]$. Thanks to Proposition 4.3.21.(iii), we have

$$\begin{aligned}\left(\mathcal{N}_{E_1^{(3)}}\right)^2 &= \left(D_0^{(2)} - 2D_1^{(2)} - 4D_2^{(2)}\right)^2 \\ &= \left(D_0^{(2)}\right)^2 + 4\left(D_1^{(2)}\right)^2 + 16\left(D_2^{(2)}\right)^2 \\ &\qquad + 4D_0^{(2)} \cdot D_1^{(2)} + 8D_0^{(2)} \cdot D_2^{(2)} + 16D_1^{(2)} \cdot D_2^{(2)} \\ &= 1 \cdot (-1) + 4 \cdot (-2) + 16 \cdot (-1) + 4 \cdot 0 + 8 \cdot 1 + 16 \cdot 1 \\ &= -1 = \left(E_1^{(3)}\right)^3.\end{aligned}$$

(iii) For $E_5^{(6)} \cong \mathbb{F}_2$ we have $\mathcal{N}_{E_5^{(6)}} = -3[H] - 2[L_\infty]$. Thanks to Proposition 4.3.21.(vi), we have

$$\begin{aligned}\left(\mathcal{N}_{E_5^{(6)}}\right)^2 &= (-3H - 2L_\infty)^2 \\ &= 9H^2 + 12H \cdot L_\infty + 4(L_\infty)^2 \\ &= 9 \cdot 0 + 12 \cdot (1) + 4 \cdot (-2) \\ &= 4 = \left(E_5^{(6)}\right)^3.\end{aligned}$$

(iv) Let us make computations for $E_0 \cong (\mathbb{F}_3)_{3a}$ in X. Here we shall show that we can make the computation also with respect to other bases of $\mathrm{Pic}(E_0)$ than the one given by Proposition 4.3.21.(vii). We have

$$\begin{aligned}\mathcal{N}_{E_0} &= -[L_\infty] - \left[H^{(3a)}\right] - 2\left[D_1^{(3a)}\right] - 4\left[D_2^{(3a)}\right] - 7\left[D_3^{(3a)}\right] \\ &= -[L_\infty] - [H] - \left[D_1^{(1)}\right] - \left[D_2^{(2)}\right] - \left[D_3^{(3a)}\right],\end{aligned}$$

and $E_0 \cdot E_0 \cdot E_0 = -4$. One can easily compute the intersection matrix for the divisors in E_0 in the basis $\{L_\infty, H, D_1^{(1)}, D_2^{(2)}, D_3^{(3a)}\}$ using Corollary 1.2.28, and obtaining

$$\begin{pmatrix} -3 & 1 & 0 & 0 & 0 \\ & 0 & 0 & 0 & 0 \\ & & -1 & 0 & 0 \\ & & & -1 & 0 \\ & & & & -1 \end{pmatrix}.$$

Hence

$$\begin{aligned} \left(\mathcal{N}_{E_0}\right)^2 &= \left(-L_\infty - H - D_1^{(1)} - D_2^{(2)} - D_3^{(3a)}\right)^2 \\ &= (L_\infty)^2 + 2L_\infty \cdot H + \left(D_1^{(1)}\right)^2 \left(D_2^{(2)}\right)^2 \left(D_3^{(3a)}\right)^2 \\ &= -3 + 2 - 1 - 1 - 1 = -4 = (E_0)^3 . \end{aligned}$$

4.4. Dynamical properties

4.4.1. Contraction to a point

In this subsection we shall give an answer to Question 4.1.12.(i) for our example, proving that there does not exist the contraction of $E = E_0 \cup \ldots \cup E_5$ to a point.

In dimension 3, a criterion such as Theorem 4.1.14 is missing, but we can replace it with stronger results in general dimension.

Definition 4.4.1. Let X be a compact complex variety, and $L \to X$ a line bundle over X. Then L is **weakly negative**, or **negative in the sense of Grauert** if there exists a continuous map $\varphi : L \to [0, +\infty)$ such that

- $\varphi^{-1}(0) = E$, where E is the null section of L,
- $\phi|_{L\setminus E}$ is C^2 and strictly plurisubharmonic.

Then the main theorem we shall need is the following criterion by Grauert.

Theorem 4.4.2. (Grauert's contraibility criterion, [32, Chapter 3, Satz 8]). *Let X be a compact complex manifold, and $E \subset X$ a divisor in X. Then there exists a contraction of E to a point if and only if the normal bundle $\mathcal{N}_E$ is weakly negative.*

In order to use the computations of Proposition 4.3.22, we shall need the following result.

Theorem 4.4.3 ([32, Chapter 3, Section 4, p. 349]). *Let X be a compact complex (non-irreducible) variety X, and $X_1, \dots, X_k$ its irreducible components. Then a line bundle $L \to X$ over X is weakly negative if and only if its restrictions $L_i := L|_{X_i}$ are weakly negative for all $i = 1, \dots, k$.*

Weakly negative line bundles are strictly related to ample line bundles.

Theorem 4.4.4. [32, Chapter 3, Section 2, Hilfssatz 1] and [45, Theorem 1.2.6 (Cartan-Serre-Grothendieck Theorem]. *Let X be a compact complex projective variety. Then a line bundle L over X is weakly negative if and only if $-L$ is ample.*

Finally, we shall need a numerical criterion for ampleness.

Theorem 4.4.5. ([45, Theorem 1.2.23 (Nakai–Moishezon–Kleiman criterion)]) *Let L be a line bundle on a compact complex projective variety X. Then L is ample if and only if*

$$L^{\dim V} \cdot V > 0 \tag{4.7}$$

for every positive-dimensional irreducible subvariety $V \subseteq X$ (including the irreducible components of X.

Corollary 4.4.6. *Let L be a line bundle on a compact complex projective irreducible surface E, such that $-L$ is effective. Let $\{D_i\}_{i=1,\dots,n}$ be a set of generators (by positive linear combinations) of effective divisors. Then L is weakly negative if and only if*

$$L \cdot D_i < 0 \tag{4.8}$$

for every $i = 1, \dots, n$.

Proof. Thanks to Theorem 4.4.4 L is weakly negative if and only if $-L$ is ample, and we can use Theorem 4.4.5(Nakai-Moishezon-Kleiman criterion). Since $-L$ is effective, we have $-L = \sum_{i=1}^{n} a_i D_i$, for suitable $a_i \in \mathbb{N}$, and if (4.8) is satisfied for every i, then

$$(-L) \cdot (-L) = -L \cdot \sum_{i=1}^{n} a_i D_i = \sum_{i=1}^{n} a_i (-L \cdot D_i) > 0,$$

so the condition (4.7) for $V = E$ is automatically satisfied. □

Theorem 4.4.7. *Let X be the Kato variety and $E_0, \dots, E_5$ the divisors described in Definition 4.2.6 . Let us set $E = \bigcup_{i=0}^{5} E_i$ the union of the supports of all these irreducible surfaces E_i for $i = 0, \dots, 5$. Then there does not exist a contraction of $E \subset X$ to a point.*

Proof. A contraction of E to a point exists if and only if we can find $r_i \in \mathbb{N}^*$ for $i = 0, \dots, 5$ such that $E := r_0E_0 + \dots + r_5E_5$ admits a contraction to a point. Thanks to Theorem 4.4.2(Grauert's contraibility criterion), this happens if and only if the normal bundle $\mathcal{N}$ of $E \subset X$ is weakly negative. Thanks to Theorem 4.4.3, this happens if and only if the restriction $\mathcal{N}_i := \mathcal{N}|_{E_i}$ is weakly negative for every $i = 0, \dots, 5$.

We want now to apply Corollary 4.4.6 to $\mathcal{N}_i$ for every i. First of all we shall compute them with respect to the divisors described in Definition 4.3.1. For each $i = 0, \dots, 5$, we have that

$$\mathcal{N}_i = r_i\mathcal{N}_{E_i} + \sum_{j\neq i} r_j\,[E_j]|_{E_i}.$$

Then, thanks to Proposition 4.3.22, we get

$$\mathcal{N}_1 = (r_0 - r_1)\left[D_0^{(2)}\right] + (r_2 - 2r_1)\left[D_1^{(2)}\right] + (r_3 - 4r_1)\left[D_2^{(2)}\right];$$

$$\mathcal{N}_2 = (r_0 + r_1 - r_2)\left[D_0^{(1)}\right] + (r_3 - 2r_2)\left[D_1^{(1)}\right];$$

$$\begin{aligned}\mathcal{N}_3 = {}&(r_0 + r_1 + r_2 - r_3)\left[D_0^{(3b)}\right] + (r_0 + r_1 + r_2 - 2r_3 + r_4)\left[D_1^{(3b)}\right]\\ &+ (2r_0 + 2r_1 + 2r_2 - 3r_3 + r_5)\left[D_2^{(3b)}\right]\\ &+ (4r_0 + 3r_1 + 3r_2 - 4r_3)\left[D_3^{(3b)}\right];\end{aligned}$$

$$\mathcal{N}_4 = (r_3 - 2r_4 + r_5)\left[D_0^{(0)}\right];$$

$$\mathcal{N}_5 = (r_0 + r_4 - 2r_5)[L_\infty] + (2r_0 + r_3 - 3r_5)[H];$$

$$\begin{aligned}\mathcal{N}_0 = {}&(-r_0 + r_5)[L_\infty] + (-r_0 + r_3)\left[H^{(3a)}\right]\\ &+ (-2r_0 + r_1 + r_3)\left[D_1^{(3a)}\right] + (-4r_0 + r_2 + 2r_3)\left[D_1^{(3a)}\right]\\ &+ (-7r_0 + 4r_3)\left[D_3^{(3a)}\right].\end{aligned}$$

Thanks to Proposition 4.3.21 we can now apply Corollary 4.4.6 to every $\mathcal{N}_i$ for $i = 0, \dots 5$.

From $\mathcal{N}_4$ we get

$$\mathcal{N}_4 \cdot D_0^{(0)} = r_3 - 2r_4 + r_5 < 0.$$

From $\mathcal{N}_2$ we get

$$\mathcal{N}_2 \cdot D_0^{(1)} = -2r_2 + r_3 < 0,$$
$$\mathcal{N}_2 \cdot D_1^{(1)} = r_0 + r_1 + r_2 - r_3 < 0.$$

From $\mathcal{N}_1$ we get

$$\mathcal{N}_1 \cdot D_0^{(2)} = -r_0 - 3r_1 + r_3 < 0,$$
$$\mathcal{N}_1 \cdot D_1^{(2)} = -2r_2 + r_3 < 0,$$
$$\mathcal{N}_1 \cdot D_2^{(2)} = r_0 + r_1 + r_2 - r_3 < 0,$$

but the last 2 inequalities had already been obtained.

From $\mathcal{N}_5$ we get

$$\mathcal{N}_5 \cdot L_\infty = r_3 - 2r_4 + r_5 < 0,$$
$$\mathcal{N}_5 \cdot H = r_0 + r_4 - 2r_5 < 0,$$

but the first inequality had already been obtained.

From $\mathcal{N}_3$ we get

$$\mathcal{N}_3 \cdot D_0^{(3b)} = 2r_0 + r_1 + r_2 - 2r_3 < 0,$$
$$\mathcal{N}_3 \cdot D_1^{(3b)} = r_3 - 2r_4 + r_5 < 0,$$
$$\mathcal{N}_3 \cdot D_2^{(3b)} = r_0 + r_4 - 2r_5 < 0,$$
$$\mathcal{N}_3 \cdot D_3^{(3b)} = -r_0 + r_5 < 0,$$

and we had already obtained the second and the third inequalities.

Finally from $\mathcal{N}_0$ we get

$$\mathcal{N}_0 \cdot L_\infty = 2r_0 + r_3 - 3r_5 < 0,$$
$$\mathcal{N}_0 \cdot H^{(3a)} = r_0 - 2r_3 + r_5 < 0,$$
$$\mathcal{N}_0 \cdot D_1^{(3a)} = -r_0 - 3r_1 + r_3 < 0,$$
$$\mathcal{N}_0 \cdot D_2^{(3a)} = -2r_2 + r_3 < 0,$$
$$\mathcal{N}_0 \cdot D_3^{(3a)} = r_0 + r_1 + r_2 - r_3 < 0,$$

where the last three inequalities had already been obtained.

Summarizing, we have to find $r_0, \ldots, r_5 > 0$ such that all the following inequalities hold:

$$\begin{aligned}
R_1 :=& -3r_1 + r_3 - r_0 < 0,\\
R_2 :=& -2r_2 + r_3 < 0,\\
R_3 :=& r_1 + r_2 - 2r_3 + 2r_0 < 0,\\
R_4 :=& r_3 - 2r_4 + r_5 < 0,\\
R_5 :=& r_4 - 2r_5 + r_0 < 0,\\
R_6 :=& r_5 - r_0 < 0,\\
R_7 :=& r_1 + r_2 - r_3 + r_0 < 0,\\
R_8 :=& r_3 - 3r_5 + 2r_0 < 0,\\
R_9 :=& -2r_3 + r_5 + r_0 < 0.
\end{aligned}$$

We can see that we can toss the 7-th and 8-th inequalities, since and $R_3 + R_4 + 2R_5 + 3R_6 = R_7$ and $R_4 + 2R_5 = R_8$, while the last one is implied by $R_3 + R_6 = R_9 + r_1 + r_2 < 0$ being $r_1, r_2 \in \mathbb{N}^*$.

Let us denote

$$V_i := \bigcap_{j \neq i} \{R_j = 0\},$$

for $i = 1, \ldots, 6$.

We shall consider as coordinates in $\mathbb{R}^6$ the sixtuple $(r_1, r_2, r_3, r_4, r_5, r_0)$. Then we get that $V_i = \mathrm{Span}\{v_i\}$, with

$$\begin{aligned}
v_1 =&(-1, 1, 2, 2, 2, 2),\\
v_2 =&(0, 0, 1, 1, 1, 1),\\
v_3 =&(0, 1, 2, 2, 2, 2),\\
v_4 =&(1, 5, 10, 7, 7, 7),\\
v_5 =&(2, 10, 20, 17, 14, 14),\\
v_6 =&(1, 5, 10, 9, 8, 7).
\end{aligned}$$

Since for every i, by direct computation, we have that $R_i(v_i) > 0$, we see that the cone given by the inequalities is generated by $\{-v_1, \ldots, -v_6\}$. In particular there are no $r_0, \ldots, r_6$ strictly positive and satisfying all the inequalities. It follows that $\mathcal{N}$ is not weakly negative, and that E cannot be contracted to a point. □

Remark 4.4.8. This example shows that there does not hold an analogous of Theorem 4.1.15 in dimension 3.

Question 4.1.12.(i) is equivalent to asking if the Alexandroff one-point compactification of a fundamental domain V of the basin of attraction U

for the base germ f_0 at 0 is a (possibly singular) complex variety. Hence Question 4.1.12.(i) can be thought as looking for a "minimal" compactification for V that is also a complex variety. While this can be always achieved just adding a point in dimension 2, the answer is not trivial in dimension 3.

4.4.2. Foliations

In this and the following subsections we shall give an answer to Question 4.1.12.(ii) for our example. Here we shall deal with the existence of a foliation $\mathcal{F}$ on our Kato variety X.

Corollary 4.4.9. *Let $f_0 : (\mathbb{C}^3, 0) \to (\mathbb{C}^3, 0)$ be the germ*

$$f_0(x, y, t) = \frac{\big(t^2(1 + yt), t^3, t^4\big)}{(1 + yt)^2 + ct^2 + xt^3}. \tag{4.4}$$

Then f_0 is holomorphically conjugated to a germ of the form

$$\widetilde{f}_0(\mathbf{x}) = \big(x_1^4, f_2(\mathbf{x}), f_3(\mathbf{x})\big) \tag{4.9}$$

where $\mathbf{x} = (x_1, x_2, x_3)$, for suitable $f_2, f_3 : \mathbb{C}^3 \to \mathbb{C}$ such that $x_1^3 \mid f_2$ and $x_1^2 \mid f_3$, by a conjugation Φ of the form

$$\mathbf{x} = \Phi(x, y, t) = \Big(t\big(1 + \phi(x, y, t)\big), y, x\Big).$$

Proof. It follows from Theorem 3.3.1 in the case $k = 1$, see also Theorem 3.2.15. □

Proposition 4.4.10. *Let X be the Kato variety and $E_0, \ldots, E_5$ the divisors described in Definition 4.2.6. Let us set $E = \bigcup_{i=0}^{5} E_i$ the union of the supports of all these irreducible surfaces E_i for $i = 0, \ldots, 5$. Then there exists a holomorphic foliation $\mathcal{F}$ of codimension 1 on X, that contains E as a leaf, and it is singular at singular points of E (*i.e.*, in $\bigcup_{i \neq j}(E_i \cap E_j)$).*

Proof. Thanks to Corollary 4.4.9, we can suppose that the base germ $f_0 = \pi \circ \sigma$ that defines X (see Subsection 4.2.3) is of the form (4.9). We can then consider in those coordinates $\mathbf{x} = (x_1, x_2, x_3)$ the 1-form $\omega = dx_1$. Since for the pull-back we have $f_0^*(\omega) = 4x_1^3 dx_1 = 4x_1^3\omega$, then ω defines a f_0-invariant holomorphic (germ of) foliation on $(\mathbb{C}^3, 0)$, and hence a foliation $\mathcal{F}$ on X, since the f_0-invariance guarantees the holomorphic gluing of leaves while defining X from $\mathcal{A}^{\pi}_{\varepsilon_-, \varepsilon_+}$ as in Subsection 4.2.3. The foliation $\mathcal{F}$ is clearly non-singular outside E, and, taking all components E_i of E as leaves, it can be extended to a foliation with singularities only in the singular set of E. □

Remark 4.4.11. Since birational maps are biholomorphisms outside a suitable divisor, admitting a foliation is clearly a birational invariant.

4.4.3. Curves and Surfaces

Before starting to study invariant curves and surfaces, we need a remark and a definition.

Remark 4.4.12. Let us consider the blow-ups $\{\pi_i\}_{i\in\mathbb{N}^*}$ we make to construct X_∞ as in Definition 4.2.9. Let us denote by $Y_i \subset X_i$ the center of $\pi_{i+1} : X_{i+1} \to X_i$ for every i.

In general, some of the Y_i can be submanifolds of dimension ≥ 1 (in our example, Y_{4+6j} and Y_{5+6j} have dimension 1 for every j), but Y_0, and hence Y_{6j} for every $j \in \mathbb{N}$, are always points.

In particular, even when we are not blowing-up points, we are interested into the geometry in a neighborhood of a single point $P_i \in Y_i$, given by the projection of a Y_{6j} for $6j \geq i$.

This remark holds not only for our example, but starting from any resolution of a strict germ, in any dimensions.

Definition 4.4.13. Let X be the Kato variety associated to the germ $f_0 : (\mathbb{C}^3, 0) \to (\mathbb{C}^3, 0)$ given by equation (4.4), with respect to the resolution $f_0 = \pi \circ \sigma$ given in Subsection 4.2.2. Let us denote by X_k the manifolds and by π_k the blow-ups as in Definition 4.2.9. We shall denote by $Y_{k-1} \subset X_{k-1}$ the center of $\pi_k : X_k \to X_{k-1}$. We shall denote by P_{6j} the only point in Y_{6j} for $j \in \mathbb{N}$, and $P_{i+6j} := \pi_{i+1+6j} \circ \ldots \circ \pi_{6+6j}(P_{6+6j}) \in Y_{i+6j}$ for $i = 1, \ldots, 5$ and $j \in \mathbb{N}$.

We shall now study the existence of invariant curves.

Lemma 4.4.14. *Let $f_0 : (\mathbb{C}^3, 0) \to (\mathbb{C}^3, 0)$ be the germ given by equation (4.4), and $f_0 = \pi \circ \sigma$ the resolution given in Subsection 4.2.2, such that (π, σ) is a Kato data. Let $\pi_k : X_k \to X_{k-1}$, $E_h^{(k)}$ be as in Definition 4.2.9, and Y_k and P_k as in Definition 4.4.13. We shall finally denote by $\sigma_j : A_j \to A_{j+1}$ the (germ of) biholomorphism induced by σ in A_j (see Remark 4.2.8), and $\sigma_0 = \sigma$.*

Let $C \not\subseteq E_0^{(0)}$ be an irreducible curve in $(\mathbb{C}^3, 0)$. Then the following conditions are equivalent.

(i) *C is invariant for f_0.*
(ii) *$\sigma(C)$ coincide with the strict transform $\widetilde{C} := \overline{\pi^{-1}(C \setminus \{0\})}$ of C through π.*
(iii) *For every $k \in \mathbb{N}$, the strict transform C_k of S through $\pi_1 \circ \ldots \circ \pi_k$ intersects $E_k^{(k)}$ in $P_k \in Y_k$, and C_{6h} is invariant for $\widetilde{f_{6h}} := \pi_{6h+1} \circ \ldots \circ \pi_{6h+6} \circ \sigma_h$ for $h \in \mathbb{N}$.*

Proof.

(i $\Rightarrow$ ii). From the hypothesis, we have that $\pi \circ \sigma(C) = C$ (being C outside $E_0^{(0)}$, the critical set of f_0); since π is a biholomorphism outside $E_6^{(6)}$, we get $\sigma(C \setminus \{0\}) = \pi^{-1}(C \setminus \{0\})$, and taking the closure we get

$$\sigma(C) = \overline{\sigma(C \setminus \{0\})} = \overline{\pi^{-1}(C \setminus \{0\})} = \widetilde{C}.$$

(ii $\Rightarrow$ iii). Let us suppose that (ii) holds; then $C_n \cap Y_n \neq \emptyset$ for $n = 1, \dots, 6$, otherwise we would have $\sigma(S) = \widetilde{S} = \emptyset$. With the same argument made in every copy A_i of A in $\widetilde{X}$ (see Subsection 4.2.3), we get the first part of the statement (see also Remark 4.4.12). Moreover

$$\begin{aligned}\pi_1 \circ \pi_{6h} \circ \widetilde{f_{6h}}(C_{6h} \setminus P_{6h-1}) &= f_0 \circ \pi_1 \circ \pi_{6h}(C_{6h} \setminus P_{6h-1}) \\ &= f_0(C \setminus \{0\}) = C \setminus \{0\} \\ &= \pi_1 \circ \pi_{6h}(C_{6h} \setminus P_{6h-1});\end{aligned}$$

taking the closure, we get the statement.

(iii $\Rightarrow$ i). Trivial.

□

Lemma 4.4.15. *Let C be a curve and E a surface in $(\mathbb{C}^3, 0)$, and $\pi : X \to (\mathbb{C}^3, 0)$ a modification over the origin. Then for the intersection number, we have*

$$C \cdot E = \widetilde{C} \cdot \pi^* E,$$

where $\widetilde{C}$ denotes the strict transform of C through π.

Proof. Let $E = \{\phi = 0\}$, and take a parametrization $C = \{\gamma(t)\}$, where $\gamma(t) = (\gamma_1(t), \gamma_2(t), \gamma_3(t))$ for suitable formal power series γ_i with $i = 1, 2, 3$. Then $\widetilde{C}$ is parametrized by a map $\widetilde{\gamma}(t)$ such that $\gamma = \pi \circ \widetilde{\gamma}$. Then directly from the definition of multiplicity of intersection we get:

$$C \cdot E = m(\phi \circ \gamma) = m(\phi \circ \pi \circ \widetilde{\gamma}) = \widetilde{C} \cdot \pi^* E,$$

where m denotes the multiplicity (with respect with the only variable t).

□

Theorem 4.4.16. *Let $f_0 : (\mathbb{C}^3, 0) \to (\mathbb{C}^3, 0)$ be the germ given by equation* (4.4)*, and $f_0 = \pi \circ \sigma$ the resolution given in Subsection* 4.2.2*, such that (π, σ) is a Kato data. Then there are no invariant curves for f_0 outside $E_0^{(0)}$.*

Proof. Let C be an invariant curve for f_0 that does not belong to $E_0^{(0)}$. Let us consider the strict transforms C_n through $\pi_1 \circ \ldots \circ \pi_n$ for every $n \in \mathbb{N}^*$. Thanks to Lemma 4.4.14 we have that $C_n \cap E_n^{(n)} = P_n \in Y_n$ for every $n \in \mathbb{N}^*$, with notations as in Definition 4.4.13. Moreover, for $n = 6h$, $P_n = Y_n$ is a free point of $E_n^{(n)}$, *i.e.*, it does not belong to any $E_k^{(n)}$ with $k < n$.

Thanks to Lemma 4.4.15, we get

$$C \cdot E_0^{(0)} = C_{6h} \cdot (\pi_1 \circ \ldots \circ \pi_{6h})^* E_0^{(0)} = 4^h C_{6h} \cdot E_{6h}^{(6h)} \geq 4^h,$$

where the last equivalence follows from the fact that C_{6h} meets the exceptional divisors only in $E_{6h}^{(6h)}$ and by direct computation.

Taking the limit for $h \to \infty$, we get a contradiction. □

Corollary 4.4.17. *Let X be the Kato variety associated to the germ $f_0 : (\mathbb{C}^3, 0) \to (\mathbb{C}^3, 0)$ given by equation* (4.4)*, with respect to the resolution $f_0 = \pi \circ \sigma$ given in Subsection* 4.2.2*. Then the only curves in X lie in $E = E_0 \cup \ldots \cup E_5$ described in Definition* 4.2.6.

Proof. It follows from the correspondence between curves on X outside E and invariant curves outside $E_0^{(0)}$ for f_0, and Theorem 4.4.16. □

Remark 4.4.18. Corollary 4.4.17 implies that there do not exist surfaces (besides $E_0, \ldots, E_5$) that contain curves (besides possibly the ones that arise from the intersection with $E_0, \ldots, E_5$). But there could still be surfaces such as for example Inoue surfaces (see [38]), with no curves inside.

Lemma 4.4.19. *Let $f_0 : (\mathbb{C}^3, 0) \to (\mathbb{C}^3, 0)$ be the germ given by equation* (4.4)*, and $f_0 = \pi \circ \sigma$ the resolution given in Subsection* 4.2.2*, such that (π, σ) is a Kato data. Let $\pi_k : X_k \to X_{k-1}$, $E_h^{(k)}$ be as in Definition* 4.2.9*, and Y_k and P_k as in Definition* 4.4.13*. We shall finally denote by $\sigma_j : A_j \to A_{j+1}$ the (germ of) biholomorphism induced by σ in A_j (see Remark* 4.2.8*), and $\sigma_0 = \sigma$.*

Let $S \neq E_0^{(0)}$ be an irreducible surface in $(\mathbb{C}^3, 0)$. Then the following conditions are equivalent.

(i) *S is invariant for f_0.*
(ii) *$\sigma(S)$ is an (open) subset of the strict transform $\widetilde{S} := \overline{\pi^{-1}(S \setminus \{0\})}$ of S through π.*
(iii) *For every $k \in \mathbb{N}$, the strict transform S_k of S through $\pi_1 \circ \ldots \circ \pi_k$ contains P_k, and S_{6h} is invariant for $\widetilde{f}_{6h} := \pi_{6h+1} \circ \ldots \circ \pi_{6h+6} \circ \sigma_h$ for $h \in \mathbb{N}$.*

Proof.
(i $\Rightarrow$ ii). From the hypothesis, we have that $\pi \circ \sigma(S) \subseteq S$; since π is a biholomorphism outside $E_6^{(6)}$, we get $\sigma(S \setminus E_0^{(0)}) \subseteq \pi^{-1}(S \setminus E_0^{(0)})$, and taking the closure we get

$$\sigma(S) = \overline{\sigma(S \setminus E_0^{(0)})} \subseteq \overline{\pi^{-1}(S \setminus E_0^{(0)})} = \widetilde{S}.$$

(ii $\Rightarrow$ iii). Let us suppose that (ii) holds; then $S_n \ni P_n$ for $n = 1, \ldots, 6$, otherwise we would have $\sigma(S) = \widetilde{S} = \emptyset$. With the same argument made in every copy A_i of A in $\widetilde{X}$ (see Subsection 4.2.3), we get the first part of the statement. Moreover

$$\begin{aligned} \pi_1 \circ \pi_{6h} \circ \widetilde{f}_{6h}(S_{6h} \setminus E_{6h}^{(6h)}) &= f_0 \circ \pi_1 \circ \pi_{6h}(S_{6h} \setminus E_{6h}^{(6h)}) \\ &= f_0(S \setminus E_0^{(0)}) \subseteq S \setminus E_0^{(0)} \\ &= \pi_1 \circ \pi_{6h}(S_{6h} \setminus E_{6h}^{(6h)}); \end{aligned}$$

taking the closure, we get the statement.
(iii $\Rightarrow$ i). Trivial. □

Remark 4.4.20. We would like to show that, in the same hypotheses of Theorem 4.4.16, the only invariant surface for f_0 is $E_0^{(0)}$. We could try to use Lemma 4.4.19 as we did with Lemma 4.4.14 for proving Theorem 4.4.16. One way to try to get a contradiction is to mimick the proof given by Dloussky (see [15, Part II, Proposition 1.10]) for an analogous of Theorem 4.4.16 for Kato surfaces. It relies on the fact that, if we have an invariant curve C, while performing the blow-ups π_k we are also desingularizing C, and this step is missing for surfaces in 3-folds.

Remark 4.4.21. We have already noticed (see Remark 4.1.11) that X is a compactification of a fundamental domain V of the basin of attraction U for f_0 to 0, obtained by "adding" 6 irreducible surfaces $E_0, \ldots, E_5$.

But one could ask what we can say about a general 3-variety Y birationally equivalent to X. In particular, we have to check what happens by performing blow-ups and blow-downs with smooth centers (see for example the weak factorization result in [9]).

But for X, we know that there are not so many curves, so we can just blow-up points in X or curves in E.

The result is that, up to modifications over points in V, a manifold Y obtained by blowing-up X is still a compactification of V.

Moreover, we cannot blow-down too many surfaces, since the ones that maybe exist outside E cannot be ruled surfaces (which have too many curves that lie inside).

Since this property on curves can be transported by blow-ups and blow-downs, we then get that a 3-variety Y that is birationally equivalent to X is still a compactification of V (up to modifications over points in V).

References

[1] M. ABATE, "An Introduction to Hyperbolic Dynamical Systems", Istituti Editoriali e Poligrafici Internazionali, Pisa, 2001.

[2] M. ABATE, *The residual index and the dynamics of holomorphic maps tangent to the identity*, Duke Math. J. (1) **107** (2001), 173–207.

[3] M. ABATE and J. RAISSY, *Formal Poincaré-Dulac renormalization for holomorphic germs*, Discrete Contin. Dyn. Syst. (5) **33** (2013), 1773–1807.

[4] M. ABATE and F. TOVENA, *Formal normal forms for holomorphic maps tangent to the identity*, Discrete Contin. Dyn. Syst. (suppl.) (2005), 1–10.

[5] A. BEAUVILLE, "Complex Algebraic Surfaces", Vol. 34 of *London Mathematical Society Student Texts*, Cambridge University Press, Cambridge, second edition, 1996. Translated from the 1978 French original by R. Barlow, with assistance from N. I. Shepherd-Barron and M. Reid.

[6] F. BERTELOOT, *Méthodes de changement d'échelles en analyse complexe*, Ann. Fac. Sci. Toulouse Math. (6) (3) **15** (2006), 427–483.

[7] XA. BUFF and J. HUBBARD, "Dynamics in one Complex Variable", Matrix Editions, Ithaca, NY, to appear.

[8] F. A. BOGOMOLOV, *Classification of surfaces of class VII0 with* $b_2 = 0$, Izv. Akad. Nauk SSSR Ser. Mat. (2) **40** (1976), 273–288, 469.

[9] L. BONAVERO, *Factorisation faible des applications birationnelles* (d'après Abramovich, Karu, Matsuki, Włodarczyk et Morelli), Astérisque (282): Exp. No. 880, vii, 1–37, 2002. Séminaire Bourbaki, Vol. 2000/2001.

[10] L. E. BÖTTCHER, *The principal laws of convergence of iterates and theri application to analysis* (russian), Izv. Kazan.Fiz.-Mat. Obshch. **14** (1904), 155–234.

[11] W. BARTH, C. PETERS and A. VAN DE VEN, "Compact Complex Surfaces", Vol. 4 of *Ergebnisse der Mathematik und ihrer Grenzgebiete (3) [Results in Mathematics and Related Areas (3)]*. Springer-Verlag, Berlin, 1984.

[12] L. CARLESON and T. W. GAMELIN, "Complex Dynamics", Universitext: Tracts in Mathematics. Springer-Verlag, New York, 1993.

[13] C. CAMACHO and P. SAD, *Invariant varieties through singularities of holomorphic vector fields*, Ann. of Math. (2) **115** (3) (1982), 579–595.

[14] F. DEGLI INNOCENTI, "On the Relations Between Discrete and Continuous Dynamics in $\mathbb{C}^2$", PhD thesis, Università di Pisa, 2007.

[15] G. DLOUSSKY, *Structure des surfaces de Kato*, Mém. Soc. Math. France (N.S.) (14), ii+120, 1984.

[16] G. DLOUSSKY and K. OELJEKLAUS, *Vector fields and foliations on compact surfaces of class* VII_0, Ann. Inst. Fourier (Grenoble) (5) **49** (1999), 1503–1545.

[17] G. DLOUSSKY and K. OELJEKLAUS, *Surfaces de la classe VII_0 et automorphismes de Hénon*, C. R. Acad. Sci. Paris Sér. I Math. (7) **328** (1999), 609–612.

[18] G. DLOUSSKY, K. OELJEKLAUS and M. TOMA, *Surfaces de la classe VII_0 admettant un champ de vecteurs*, Comment. Math. Helv. (2) **75** (2000), 255–270.

[19] G. DLOUSSKY, K. OELJEKLAUS and M. TOMA, *Surfaces de la classe VII_0 admettant un champ de vecteurs*, II, Comment. Math. Helv. (4) **76** (2001), 640–664.

[20] G. DLOUSSKY, K. OELJEKLAUS and M. TOMA, *Class* VII_0 *surfaces with b_2 curves*, Tohoku Math. J. (2) **55** (2003), 283–309.

[21] J. ÉCALLE, "Les fonctions résurgentes", Tome I, volume 5 of Publications Mathématiques d'Orsay 81 [Mathematical Publications of Orsay 81], Université de Paris-Sud Département de Mathématique, Orsay, 1981. "Les algèbres de fonctions résurgentes" [The algebras of resurgent functions], With an English foreword.

[22] J. ÉCALLE, "Les fonctions résurgentes", Tome II, volume 6 of Publications Mathématiques d'Orsay 81 [Mathematical Publications of Orsay 81], Université de Paris-Sud Département de Mathématique, Orsay, 1981. "Les fonctions résurgentes appliquées à l'itération", [Resurgent functions applied to iteration].

[23] J.ÉCALLE, "Les fonctions résurgentes", Tome III, volume 85 of Publications Mathématiques d'Orsay [Mathematical Publications of Orsay], Université de Paris-Sud, Département de Mathé-

matiques, Orsay, 1985. "L'équation du pont et la classification analytique des objects locaux", [The bridge equation and analytic classification of local objects].

[24] P. FATOU, *Sur les équations fonctionnelles*, Bull. Soc. Math. France **47** (1919), 161–271.

[25] C. FAVRE, *Classification of 2-dimensional contracting rigid germs and Kato surfaces*, I, J. Math. Pures Appl. (9) **79** (2000), 475–514.

[26] C. FAVRE and M. JONSSON, "The Valuative Tree", Vol. 1853 of Lecture Notes in Mathematics, Springer-Verlag, Berlin, 2004.

[27] C. FAVRE and M. JONSSON, *Eigenvaluations*, Ann. Sci. École Norm. Sup. (4) **40** (2007), 309–349.

[28] H. M. FARKAS and I. KRA, "Riemann surfaces", Vol. 71 of Graduate Texts in Mathematics, Springer-Verlag, New York, 1980.

[29] S. FRIEDLAND and J. MILNOR, *Dynamical properties of plane polynomial automorphisms*, Ergodic Theory Dynam. Systems (1) **9** (1989), 67–99.

[30] J. E. FORNÆSS and H. WU, *Classification of degree* 2 *polynomial automorphisms of* $\mathbf{C}^3$, Publ. Mat. (1) **42** (1998), 195–210.

[31] P. GRIFFITHS and J. HARRIS, "Principles of Algebraic Geometry", Wiley-Interscience [John Wiley & Sons], New York, 1978, Pure and Applied Mathematics.

[32] H. GRAUERT, *Über Modifikationen und exzeptionelle analytische Mengen*, Math. Ann. **146** (1962), 331–368.

[33] M. HAKIM, *Attracting domains for semi-attractive transformations of* $\mathbf{C}^p$, Publ. Mat. (2) **38** (1994), 479–499.

[34] M. HAKIM, *Analytic transformations of* $(\mathbf{C}^p, 0)$ *tangent to the identity*, Duke Math. J. (2) **92** (1998), 403–428.

[35] R. HARTSHORNE, "Algebraic Geometry", Springer-Verlag, New York, 1977. Graduate Texts in Mathematics, No. 52.

[36] A. HATCHER, "Algebraic Topology", Cambridge University Press, Cambridge, 2002.

[37] J. H. HUBBARD and R. W. OBERSTE-VORTH, Hénon mappings in the complex domain. I. The global topology of dynamical space. Inst. Hautes Études Sci. Publ. Math. (79) (1994), 5–46.

[38] M. INOUE, *On surfaces of Class* VII_0, Invent. Math. **24** (1974), 269–310.

[39] V. A. ISKOVSKIKH and I. R. SHAFAREVICH, *Algebraic surfaces* [MR1060325 (91f:14029)], In: "Algebraic Geometry", II, Vol. 35 of Encyclopaedia Math. Sci., Springer, Berlin, 1996, 127–262.

[40] M. KATO, *Compact complex manifolds containing "global" spherical shells*, I, In: "Proceedings of the International Symposium on

Algebraic Geometry" (Kyoto Univ., Kyoto, 1977), Tokyo, 1978. Kinokuniya Book Store, 45–84.

[41] M. KATO, *On compact complex 3-folds with lines*, Japan. J. Math. (N.S.) (1) **11** (1985), 1–58.

[42] K. KODAIRA, *On the structure of compact complex analytic surfaces*, I, Amer. J. Math. **86** (1964), 751–798.

[43] K. KODAIRA, *On the structure of compact complex analytic surfaces*, II, Amer. J. Math. **88** (1966), 682–721.

[44] G. KŒNIGS, *Recherches sur les intégrales de certaines équations fonctionnelles*, Ann. Sci. École Norm. Sup. (3) **1** (1884), 3–41.

[45] R. LAZARSFELD, "Positivity in Algebraic Geometry", I, Vol. 48 of Ergebnisse der Mathematik und ihrer Grenzgebiete. 3. Folge. A. Series of Modern Surveys in Mathematics [Results in Mathematics and Related Areas. 3rd Series. A Series of Modern Surveys in Mathematics], Springer-Verlag, Berlin, 2004. Classical setting: line bundles and linear series.

[46] L. LEAU, *Étude sur les équations fonctionnelles à une ou à plusieurs variables*, Ann. Fac. Sci. Toulouse Sci. Math. Sci. Phys. (2) **11** (1897), E1–E24.

[47] L. LEAU, *Étude sur les équations fonctionnelles à une ou à plusieurs variables*, Ann. Fac. Sci. Toulouse Sci. Math. Sci. Phys. (3) **11** (1897), E25–E110.

[48] J. LI, S.-T. YAU and F. ZHENG, *On projectively flat Hermitian manifolds*, Comm. Anal. Geom. (1) **2** (1994), 103–109.

[49] K. MAEGAWA, *Classification of quadratic polynomial automorphisms of $\mathbb{C}^3$ from a dynamical point of view*, Indiana Univ. Math. J. (2) **50** (2001), 935–951.

[50] S. MARMI, "An Introduction to Small Divisors Problems", Istituti Editoriali e Poligrafici Internazionali, Pisa, 2000.

[51] J. MILNOR, "Dynamics in one Complex Variable", Vol. 160 of Annals of Mathematics Studies, Princeton University Press, Princeton, NJ, third edition, 2006.

[52] K. OELJEKLAUS and J. RENAUD, *Compact complex threefolds of class L associated to polynomial automorphisms of $\mathbb{C}^3$*, Publ. Mat. (2) **50** (2006), 401–411.

[53] J. RIVERA-LETELIER, *Sur la structure des ensembles de Fatou p-adiques*, Preprint available at www.math.sunysb.edu/ rivera.

[54] J.-P. ROSAY and W. RUDIN, *Holomorphic maps from $\mathbf{C}^n$ to $\mathbf{C}^n$*, Trans. Amer. Math. Soc. (1) **310** (1988), 47–86.

[55] A. SEIDENBERG, *Reduction of singularities of the differential equation $A\,dy = B\,dx$*, Amer. J. Math. **90** (1968), 248–269.

[56] I. R. SHAFAREVICH, "Basic Algebraic Geometry", 1, Springer-Verlag, Berlin, second edition, 1994. Varieties in projective space, Translated from the 1988 Russian edition and with notes by Miles Reid.

[57] N. SIBONY, *Dynamique des applications rationnelles de* $\mathbf{P}^k$, In: "Dynamique et géométrie complexes" (Lyon, 1997), Vol. 8 of Panor. Synthèses, pages ix–x, xi–xii, 97–185. Soc. Math. France, Paris, 1999.

[58] M. SPIVAKOVSKY, *Valuations in function fields of surfaces*, Amer. J. Math. (1) **112** (1990), 107–156.

[59] S. STERNBERG, *Local contractions and a theorem of Poincaré*, Amer. J. Math. **79** (1957), 809–824.

[60] A. D. TELEMAN, *Projectively flat surfaces and Bogomolov's theorem on class* VII_0 *surfaces*, Internat. J. Math. (2) **5** (1994), 253–264.

[61] S. M. VORONIN, *Analytic classification of germs of conformal mappings* $(\mathbf{C}, 0) \to (\mathbf{C}, 0)$, Funktsional. Anal. i Prilozhen. (1) **15** (1981), 1–17, 96.

[62] R. O. WELLS, JR., "Differential Analysis on Complex Manifolds", Vol. 65 of Graduate Texts in Mathematics, Springer, New York, third edition, 2008. With a new appendix by Oscar Garcia-Prada.

[63] O. ZARISKI and P. SAMUEL, "Commutative Algebra", Vol. 1, Springer-Verlag, New York, 1975. With the cooperation of I. S. Cohen, Corrected reprinting of the 1958 edition, Graduate Texts in Mathematics, No. 28.

[64] O. ZARISKI and P. SAMUEL, "Commutative Algebra", Vol. II, Springer-Verlag, New York, 1975. Reprint of the 1960 edition, Graduate Texts in Mathematics, Vol. 29.

Index

THESES

This series gathers a selection of outstanding Ph.D. theses defended at the Scuola Normale Superiore since 1992.

Published volumes

1. F. COSTANTINO, *Shadows and Branched Shadows of* 3 *and* 4*-Manifolds*, 2005. ISBN 88-7642-154-8
2. S. FRANCAVIGLIA, *Hyperbolicity Equations for Cusped* 3*-Manifolds and Volume-Rigidity of Representations*, 2005. ISBN 88-7642-167-x
3. E. SINIBALDI, *Implicit Preconditioned Numerical Schemes for the Simulation of Three-Dimensional Barotropic Flows*, 2007. ISBN 978-88-7642-310-9
4. F. SANTAMBROGIO, *Variational Problems in Transport Theory with Mass Concentration*, 2007. ISBN 978-88-7642-312-3
5. M. R. BAKHTIARI, *Quantum Gases in Quasi-One-Dimensional Arrays*, 2007. ISBN 978-88-7642-319-2
6. T. SERVI, *On the First-Order Theory of Real Exponentiation*, 2008. ISBN 978-88-7642-325-3
7. D. VITTONE, *Submanifolds in Carnot Groups*, 2008. ISBN 978-88-7642-327-7
8. A. FIGALLI, *Optimal Transportation and Action-Minimizing Measures*, 2008. ISBN 978-88-7642-330-7
9. A. SARACCO, *Extension Problems in Complex and CR-Geometry*, 2008. ISBN 978-88-7642-338-3
10. L. MANCA, *Kolmogorov Operators in Spaces of Continuous Functions and Equations for Measures*, 2008. ISBN 978-88-7642-336-9

11. M. LELLI, *Solution Structure and Solution Dynamics in Chiral Ytterbium(III) Complexes*, 2009. ISBN 978-88-7642-349-9
12. G. CRIPPA, *The Flow Associated to Weakly Differentiable Vector Fields*, 2009. ISBN 978-88-7642-340-6
13. F. CALLEGARO, *Cohomology of Finite and Affine Type Artin Groups over Abelian Representations*, 2009. ISBN 978-88-7642-345-1
14. G. DELLA SALA, *Geometric Properties of Non-compact CR Manifolds*, 2009. ISBN 978-88-7642-348-2
15. P. BOITO, *Structured Matrix Based Methods for Approximate Polynomial GCD*, 2011. ISBN: 978-88-7642-380-2; e-ISBN: 978-88-7642-381-9
16. F. POLONI, *Algorithms for Quadratic Matrix and Vector Equations*, 2011. ISBN: 978-88-7642-383-3; e-ISBN: 978-88-7642-384-0
17. G. DE PHILIPPIS, *Regularity of Optimal Transport Maps and Applications*, 2013. ISBN: 978-88-7642-456-4; e-ISBN: 978-88-7642-458-8
18. G. PETRUCCIANI, *The Search for the Higgs Boson at CMS*, 2013. ISBN: 978-88-7642-481-6; e-ISBN: 978-88-7642-482-3
19. B. VELICHKOV, *Existence and Regularity Results for Some Shape Optimization Problems*, 2015. ISBN: 978-88-7642-526-4; e-ISBN: 978-88-7642-527-1
20. M. RUGGIERO, *Rigid Germs, the Valuative Tree, and Applications to Kato Varieties*, 2015. ISBN: 978-88-7642-558-5 e-ISBN: 978-88-7642-559-2

Volumes published earlier

H. Y. FUJITA, *Equations de Navier-Stokes stochastiques non homogènes et applications*, 1992.

G. GAMBERINI, *The minimal supersymmetric standard model and its phenomenological implications*, 1993. ISBN 978-88-7642-274-4

C. DE FABRITIIS, *Actions of Holomorphic Maps on Spaces of Holomorphic Functions*, 1994. ISBN 978-88-7642-275-1

C. PETRONIO, *Standard Spines and 3-Manifolds*, 1995. ISBN 978-88-7642-256-0

I. DAMIANI, *Untwisted Affine Quantum Algebras: the Highest Coefficient of* det H_η *and the Center at Odd Roots of 1*, 1996. ISBN 978-88-7642-285-0

M. MANETTI, *Degenerations of Algebraic Surfaces and Applications to Moduli Problems*, 1996. ISBN 978-88-7642-277-5

F. CEI, *Search for Neutrinos from Stellar Gravitational Collapse with the MACRO Experiment at Gran Sasso*, 1996. ISBN 978-88-7642-284-3

A. SHLAPUNOV, *Green's Integrals and Their Applications to Elliptic Systems*, 1996. ISBN 978-88-7642-270-6

R. TAURASO, *Periodic Points for Expanding Maps and for Their Extensions*, 1996. ISBN 978-88-7642-271-3

Y. Bozzi, *A study on the activity-dependent expression of neurotrophic factors in the rat visual system*, 1997. ISBN 978-88-7642-272-0

M. L. Chiofalo, *Screening effects in bipolaron theory and high-temperature superconductivity*, 1997. ISBN 978-88-7642-279-9

D. M. Carlucci, *On Spin Glass Theory Beyond Mean Field*, 1998. ISBN 978-88-7642-276-8

G. Lenzi, *The MU-calculus and the Hierarchy Problem*, 1998. ISBN 978-88-7642-283-6

R. Scognamillo, *Principal G-bundles and abelian varieties: the Hitchin system*, 1998. ISBN 978-88-7642-281-2

G. Ascoli, *Biochemical and spectroscopic characterization of CP20, a protein involved in synaptic plasticity mechanism*, 1998. ISBN 978-88-7642-273-7

F. Pistolesi, *Evolution from BCS Superconductivity to Bose-Einstein Condensation and Infrared Behavior of the Bosonic Limit*, 1998. ISBN 978-88-7642-282-9

L. Pilo, *Chern-Simons Field Theory and Invariants of 3-Manifolds*, 1999. ISBN 978-88-7642-278-2

P. Aschieri, *On the Geometry of Inhomogeneous Quantum Groups*, 1999. ISBN 978-88-7642-261-4

S. Conti, *Ground state properties and excitation spectrum of correlated electron systems*, 1999. ISBN 978-88-7642-269-0

G. Gaiffi, *De Concini-Procesi models of arrangements and symmetric group actions*, 1999. ISBN 978-88-7642-289-8

N. Donato, *Search for neutrino oscillations in a long baseline experiment at the Chooz nuclear reactors*, 1999. ISBN 978-88-7642-288-1

R. Chirivì, *LS algebras and Schubert varieties*, 2003. ISBN 978-88-7642-287-4

V. Magnani, *Elements of Geometric Measure Theory on Sub-Riemannian Groups*, 2003. ISBN 88-7642-152-1

F. M. Rossi, *A Study on Nerve Growth Factor (NGF) Receptor Expression in the Rat Visual Cortex: Possible Sites and Mechanisms of NGF Action in Cortical Plasticity*, 2004. ISBN 978-88-7642-280-5

G. Pintacuda, *NMR and NIR-CD of Lanthanide Complexes*, 2004. ISBN 88-7642-143-2

Fotocomposizione "CompoMat", Loc. Braccone, 02040 Configni (RI) Italia
Finito di stampare nel mese di dicembre 2015
dalla CSR srl, Via di Salone, 31/c, 00131 Roma, Italia